ECCION DE GEOGRAFIA - BIBLIOTECA
AF872883

$ 7.00
Geografía
R.G.-H.-55.605
R. 16.804
22.4/189

FAO TECHNICAL GUIDELINES FOR RESPONSIBLE FISHERIES

4

SECCION DE GEOGRAFIA - BIBLIOTECA

FISHERIES MANAGEMENT

EJEMPLAR EXPURGADO
H Ca. 475/015

FOOD AND AGRICULTURE ORGANIZATION OF THE UNITED NATIONS
Rome, 1997

UNIVERSIDAD DE SEVILLA
Fac. G.ª e Historia-Biblioteca

The designations employed and the presentation of material in this publication do not imply the expression of any opinion whatsoever on the part of the Food and Agriculture Organization of the United Nations concerning the legal status of any country, territory, city or area or of its authorities, or concerning the delimitation of its frontiers or boundaries.

M-43
ISBN 92-5-103962-3

All rights reserved. No part of this publication may be reproduced, stored in a retrieval system, or transmitted in any form or by any means, electronic, mechanical, photocopying or otherwise, without the prior permission of the copyright owner. Applications for such permission, with a statement of the purpose and extent of the reproduction, should be addressed to the Director, Information Division, Food and Agriculture Organization of the United Nations, Viale delle Terme di Caracalla, 00100 Rome, Italy.

© **FAO 1997**

PREPARATION OF THIS DOCUMENT

The present Guidelines have been completed by the Fishery Resources Division and the Fishery Policy and Planning Division, Fisheries Department, with the collaboration of the FAO Legal Office, on the basis of inputs provided by many experts from outside and inside FAO over the last two years, including staff of government laboratories, universities and non-governmental organizations. We wish to thank in particular those who attended the Expert Consultation in New Zealand in January 1995, of whom we must mention Mr R. Allen (Chairman, New Zealand), as well as Mr M. Lodge (U.K.) and Mr M.P. Sissenwine (U.S.A.), Chairmen of the two working groups. Several experts also deserve individual mention: Mr J.F. Caddy (FAO Fisheries Department), who coordinated the work, Mr B. Pollock (Australia) and Mr J.J. Maguire (Canada), who subsequently helped with redrafting. Mr J. Kurien (India) and Mr T. Kenchington (Canada) also provided written inputs. Among FAO staff, D. Doulman, S. Garcia, R. Grainger, C. Newton, R. Welcomme, U. Wijkström and R. Willmann assisted in various ways. The final draft was prepared by A. Bonzon and K. Cochrane.

It has to be stressed that these Guidelines have no formal legal status. They are intended to provide support for the implementation of the Code of Conduct for Responsible Fisheries. Furthermore, in order to present the management process in all its complexity and diversity, the wording and structure of these Guidelines do not follow strictly the language and the structure of the Code. Therefore, any eventual differences in the terminology employed should not be understood as intending reinterpretation of the Code. Finally, it should be remembered that, since the Guidelines are intended to be flexible and capable of evolving as circumstances change, or as new information becomes available, they may be further revised and complemented by other guidelines, notes, etc., on specific issues.

Distribution:

All FAO Members and Associate Members
Interested Nations and International Organizations
FAO Fisheries Department
FAO Fisheries Officers in FAO Regional Offices
Interested Non-Governmental Organizations

UNIVERSIDAD DE SEVILLA
Fac. G.ª e Historia-Biblioteca

FAO Fishery Resources Division and Fishery Policy and Planning Division.
Fisheries management.
FAO Technical Guidelines for Responsible Fisheries. No. 4. Rome, FAO. 1997. 82p.

ABSTRACT

These Guidelines have been produced to support the implementation of Article 7 of the Code of Conduct for Responsible Fisheries, with some reference to Article 12. They are addressed primarily to the decision-makers within fisheries management authorities and other interest groups, including fishing companies, fishers' organizations, concerned non-governmental organizations and others.

The Guidelines provide a background to the need for fisheries management and an introduction to the activities encompassed by fisheries management. They introduce the major constraints experienced in fisheries and fisheries management and some of the fundamental concepts related to these. Biological, environmental, technological, socio-cultural and economic constraints and concepts are examined.

Information is fundamental to responsible fisheries management and these Guidelines put emphasis on the range of data required for informed decision-making and examine aspects of the collection and interpretation of these data. Data are discussed in terms of three suggested scales in fisheries management: fisheries policy and development planning, formulation of management plans and implementation of management action.

The range of possible management actions is outlined. This includes technical measures, such as gear restrictions, and more direct approaches in the form of direct catch limitation or effort limitation. The problems associated with open access fisheries are explained and comments made on the means to limit access and obstacles which may be encountered in this process.

Finally, the Guidelines examine the management process. This section covers the process of agreeing on a management plan for a fishery, including the need for consultation and, where appropriate, cooperative decision-making. The need for periodic review of management plans is stressed. The importance of an effective legal framework, institutional and administrative structures and monitoring control and surveillance are described.

CONTENTS[1]

[1]*Throughout the text, references to the Code of Conduct for Responsible Fisheries are shown in italics.* These references are made in broad terms and do not strictly follow the order nor do they cover all occasions on which issues dealt with by the Code of Conduct are referred to. References to other sections within these Guidelines are shown in normal font.

Page

BACKGROUND

1. From ancient times, fishing has been a major source of food for humanity and a provider of employment and economic benefits to those engaged in this activity. However, with increased knowledge and the dynamic development of fisheries, it was realized that living aquatic resources, although renewable, are not infinite and need to be properly managed, if their contribution to the nutritional, economic and social well-being of the growing world's population was to be sustained.

2. The adoption in 1982 of the United Nations Convention on the Law of the Sea provided a new framework for the better management of marine resources. The new legal regime of the oceans gave coastal States rights and responsibilities for the management and use of fishery resources within their EEZs, which embrace some 90 percent of the world's marine fisheries.

3. In recent years, world fisheries have become a dynamically developing sector of the food industry, and many States have striven to take advantage of their new opportunities by investing in modern fishing fleets and processing factories in response to growing international demand for fish and fishery products. It became clear, however, that many fisheries resources could not sustain an often uncontrolled increase of exploitation.

4. Clear signs of over-exploitation of important fish stocks, modifications of ecosystems, significant economic losses, and international conflicts on management and fish trade threatened the long-term sustainability of fisheries and the contribution of fisheries to food supply. Therefore, the Nineteenth Session of the FAO Committee on Fisheries (COFI), held in March 1991, recommended that new approaches to fisheries management embracing conservation and environmental, as well as social and economic, considerations were urgently needed. FAO was asked to develop the concept of responsible fisheries and elaborate a Code of Conduct to foster its application.

5. Subsequently, the Government of Mexico, in collaboration with FAO, organized an International Conference on Responsible Fishing in Cancún in May 1992. The Declaration of Cancún endorsed at that Conference was brought to the attention of the UNCED Summit in Rio de Janeiro, Brazil, in June 1992, which supported the preparation of a Code of Conduct for Responsible Fisheries. The

FAO Technical Consultation on High Seas Fishing, held in September 1992, further recommended the elaboration of a Code to address the issues regarding high seas fisheries.

6. The One Hundred and Second Session of the FAO Council, held in November 1992, discussed the elaboration of the Code, recommending that priority be given to high seas issues and requested that proposals for the Code be presented to the 1993 session of the Committee on Fisheries.

7. The Twentieth Session of COFI, held in March 1993, examined in general the proposed framework and content for such a Code, including the elaboration of guidelines, and endorsed a time frame for the further elaboration of the Code. It also requested FAO to prepare, on a "fast track" basis, as part of the Code, proposals to prevent reflagging of fishing vessels which affect conservation and management measures on the high seas. This resulted in the FAO Conference, at its Twenty-seventh Session in November 1993, adopting the Agreement to Promote Compliance with International Conservation and Management Measures by Fishing Vessels on the High Seas, which, according to FAO Conference Resolution 15/93, forms an integral part of the Code.

8. The Code was formulated so as to be interpreted and applied in conformity with the relevant rules of international law, as reflected in the United Nations Convention on the Law of the Sea, 1982, as well as with the Agreement for the Implementation of the Provisions of the United Nations Convention on the Law of the Sea of 10 December 1982 Relating to the Conservation and Management of Straddling Fish Stocks and Highly Migratory Fish Stocks, 1995, and in the light of, *inter alia*, the 1992 Declaration of Cancún and the 1992 Rio Declaration on Environment and Development, in particular Chapter 17 of Agenda 21.

9. The development of the Code was carried out by FAO in consultation and collaboration with relevant United Nations Agencies and other international organizations, including non-governmental organizations.

10. The Code of Conduct consists of five introductory articles: Nature and Scope; Objectives; Relationship with Other International Instruments; Implementation, Monitoring and Updating and Special Requirements of Developing Countries. These introductory articles are followed by an article on General Principles, which precedes the six thematic articles on Fisheries Management, Fishing Operations, Aquaculture Development, Integration of Fisheries into Coastal Area Management, Post-Harvest Practices and Trade, and Fisheries Research. As already mentioned, the Agreement to Promote Compliance with International

Conservation and Management Measures by Fishing Vessels on the High Seas forms an integral part of the Code.

11. The Code is voluntary. However, certain parts of it are based on relevant rules of international law, as reflected in the United Nations Convention on the Law of the Sea of 10 December 1982. The Code also contains provisions that may be or have already been given binding effect by means of other obligatory legal instruments amongst the Parties, such as the Agreement to Promote Compliance with Conservation and Management Measures by Fishing Vessels on the High Seas, 1993.

12. The Twenty-eighth Session of the Conference in Resolution 4/95 adopted the Code of Conduct for Responsible Fisheries on 31 October 1995. The same Resolution requested FAO *inter alia* to elaborate as appropriate technical guidelines in support of the implementation of the Code in collaboration with members and interested relevant organizations.

INTRODUCTORY NOTE TO THE GUIDELINES FOR FISHERIES MANAGEMENT

These guidelines are addressed primarily to the decision-makers within fisheries management authorities and interest groups, including fishing companies, fishers' organizations, concerned non-governmental organizations (NGO) and others.

The formulation of guidelines in support of the implementation of Article 7 (Fisheries Management) of the Code of Conduct for Responsible Fisheries has been a demanding task, and one that can only be expected to provide an overall view of the key issues concerned. In fact, beginning with the Expert Consultation on this subject held in Auckland, New Zealand, 23-27 January 1995 (*FAO Fish.Rep.*, 519), the original draft annexed to that report has gone through a number of revisions, reflecting inputs from a large number of experts in many different fields within FAO and outside, as well as from intergovernmental and non-governmental organizations.

The key problems faced by the Organization in preparing the guidelines and by the experts who have provided input to this document fall into six main categories:

1. The highly diverse approaches to fisheries management currently adopted by States throughout the world involved in fisheries reflect their traditions, infrastructure and ecological and geographical situation, and their varied approaches to the question of individual rights to exploit national resources.

2. Management guidelines must be truly interdisciplinary, and their formulation requires agreement, not only on the technical details by local experts on resources, economics, and sociology, but also on common paradigms and objectives, before the details of a management system can be formulated. To a significant extent, this process must be accomplished in the local context and cannot be provided in the form of a 'recipe'.

3. The guidelines must have something relevant to say at all levels of the fishery process, from international commissions, national governments and local communities to fishing enterprises and fishermen.

4. The Code is intended to be general, covering freshwater, coastal and estuarine to high seas fisheries, and from small scale to sophisticated commercial fisheries. Although there are some global themes which we have

tried to emphasize, a number of the relevant considerations for specific categories differ in each case.

5. A lack of confidence in some of the methods of fisheries management currently more widely used has developed. This, in practical terms, stems from the seriously depleted state of many world fisheries that have used these methods (see, for example, *FAO Fish.Tech.Pap.* 335, 'Review of the state of the world marine fishery resources', and subsequent FAO fishery reviews). This does not mean that the methodologies in question cannot be effective in certain circumstances but does means that standard recipes cannot be offered and should not be accepted without careful consideration of their relevance in the local situation.

6 The current period is, in fact, one of considerable experimentation with new approaches to fisheries management, including, *inter alia*, technical measures, economic and social tools for ensuring inter-generational equity, 'top-down' and 'bottom-up' approaches, input and output control mechanisms and management frameworks involving management in partnership between the State and fishers or their communities. This search for appropriate new approaches, without abandoning the traditional measures where they have proved effective, is to be actively encouraged. Similarly, searches for more appropriate combinations of established management tools may well be fruitful in many cases. It is therefore to be expected that, in future, further additions to, revisions of or commentaries on these guidelines will be issued by FAO and others which emphasize and elaborate in much more detail specific management approaches in particular situations for specific resources.

All of these considerations make it impossible for this document to provide a single prescription of optimal management in the case of a given fishery. However, it is hoped that, in conjunction with the 'Guidelines on the precautionary approach to capture fisheries and species introductions' (*FAO Fish.Tech.Pap.*, 350/1, reissued as *FAO Technical Guidelines for Responsible Fisheries*. No.2. Rome, FAO. 1996. 54p.), the 'Reference points for fisheries management' (*FAO Fish.Tech.Pap.*, 347) and the other guidelines for the Code issued on 'Integration of fisheries into coastal area management' and on 'Fishing operations' (both of which are being published in the same series as the present one), these guidelines will help to focus an informed search for a management framework appropriate to the particular circumstance being considered.

Moritaka Hayashi
Assistant Director-General (Fisheries Department)

1. INTRODUCTION

1.1 The Need for Fisheries Management

Fisheries have substantial social and economic importance. It is estimated that 12.5 million people are employed in activities related to fishing and the value of fish traded internationally has been estimated at US$ 40 billion per annum for the early nineties. The total production from capture fisheries and aquaculture during the same period reached and oscillated around a total mass of 100 million tons.

However, at present, a large proportion of the world's exploited fish stocks are fully exploited, over-exploited, depleted or in need of recovery and many are affected by environmental degradation, particularly in the inland and coastal areas. Major ecological damage, which may not always be reversible, and economic waste are already evident in many cases.

New technological developments, such as geographical positioning systems (GPS), radar, echo-sounders, more powerful vessels and improved processing methods (e.g. surimi) continue to enhance the ability of fishers to exploit more living resources more intensively, potentially increasing the severity of the problem.

The existing status of the world's living aquatic resources is largely a result of a failure of the present process of fisheries governance to achieve responsible and effective management of fisheries in most countries. Fishers, fisheries management authorities and fisheries scientists, as well as those responsible for indirect impacts such as environmental degradation, must accept shared responsibility for the existing unsatisfactory status of the world's fisheries and living aquatic resources. It is the States' responsibility to ensure that joint measures are taken to reverse these trends. What follows highlights the actions necessary to implement responsible fisheries management.

It is important for fisheries managers to realize that, when resources are being over-exploited or exploited in an irresponsible manner, a failure to act will have negative consequences in the future. Reducing fish stocks to biologically and ecologically harmful levels will result in a loss of potential benefits as food, income, employment and others, both immediately and in the long term. A very low level of any stock is likely to have negative impacts on other dependent stocks, and the losses may extend beyond the immediately affected stock. It cannot automatically be assumed that, in such cases, a relaxation of fishing pressure will lead to a full or immediate recovery of the stock and associated ecosystem. In some cases losses may be long-lasting or even permanent.

1.2 The Fisheries Management Process

There are no clear and generally accepted definitions of fisheries management. A working definition, for the purposes of this document, may be taken as:

> The integrated process of information gathering, analysis, planning, consultation, decision-making, allocation of resources and formulation and implementation, with enforcement as necessary, of regulations or rules which govern fisheries activities in order to ensure the continued productivity of the resources and accomplishment of other fisheries objectives.

Fisheries management entails a complex and wide-embracing set of tasks, aimed at ensuring that the optimal benefits are obtained for the local users, State or region from the sustainable utilization of the living aquatic resources to which they have access. While fisheries management draws on fisheries research and analysis and, sometimes, institutionalized processes of elaboration of advice, it should not be confused with them; it encompasses but goes beyond them. From the working definition above, fisheries management can be taken to include the following:

(i) Setting policies and objectives for each fishery or stock to be managed, taking into account the biological characteristics of the stock, the nature of existing or potential fisheries and other activities related to or impacting the stock and the potential economic and social contribution of the fishery to national or local needs and goals.

(ii) Determining and implementing the actions necessary to enable the management authorities, the fishers and other interest groups, to work towards the identified objectives. This task should be done in consultation with all interest groups. The actions required will include: developing and implementing management plans for all managed stocks; ensuring that the stock or stocks, the ecosystems in which they occur and their environment are maintained in a productive state; collecting and analyzing the biological and fishery data necessary for assessment, monitoring, control and surveillance; adoption and promulgation of appropriate and effective laws and regulations necessary to achieve the objectives, and ensuring that fishers comply with them to achieve the objectives.

(iii) Consulting and negotiating with users or interest groups concerned with resources and from areas not directly related to fishery activities but which impact on fisheries. Examples would include groups engaged in activities in a river or lake basin or the coastal zone which impact on fisheries. The management authority needs to ensure that the interests of fisheries are appropriately considered and catered for in planning and integration of such activities.

(iv) In consultation with the users, regularly reviewing the management objectives and measures to ensure they are still appropriate and effective.

(v) Reporting to Governments, users and the public on the state of resources and management performance.

In the Code of Conduct the primary responsibility for overseeing the fisheries management process is vested essentially with fisheries arrangements and organizations. To be operational and allow for adequate governance, those arrangements and organizations should be determined and integrated into institutional support structures, i.e the fisheries management institutions. In this document, the fisheries management institutions have been arbitrarily aggregated into two main categories: the fisheries management authority and the interested parties.

A fisheries management authority is defined very broadly in this document in order to focus attention on the fisheries management process itself rather than focusing attention on the otherwise important juridical distinctions between the very wide range of bodies (both national and international) embraced by the term as it is defined here.

A fisheries management authority accordingly is defined to include the legal entity which has been assigned by a State or States with a mandate to perform certain specified fisheries management functions. In national systems, including federal systems, a fisheries management authority would usually take the form of a ministry, a department within a ministry (e.g. agriculture) or an agency. A fisheries management authority may also be international in character and include a fisheries management organization or arrangement either sub-regional, regional or global. No assumption is made in this definition as to whether the body in question is governmental, parastatal or private.

The term interested party (or interest group) refers to any party which has been accepted by the State or States or by the management authority on behalf of the State or States as having a legitimate interest in the fisheries resources being managed.

1.3 Biological and Environmental Concepts and Constraints (*7.1.1*)

1.3.1 Resource constraints

(i) Living populations or stocks are capable of growth in abundance and biomass, but only up to a certain limit. The limits to growth are determined by the current size of the population in relation to its average abundance in the unexploited state, and by the environment in which the stock occurs. The maintenance of a stock at productive levels requires an adequate abundance of reproductively mature adults, the spawners, and suitable critical environments for successfully passing through the different stages in the life history. However, particularly as a result of variability in the environment, the growth of a stock from year to year is usually highly variable.

(ii) The potential productivity of stocks is best understood through scientific analysis, based on agreed concepts, using standard methodologies to produce reproducible and comparable results. However, where the capacity is not available for this, for example in some traditional coastal communities, some estimates may be obtainable from empirical observation on historical catch levels. Determining the existing status of a stock, and the potential yield from it under different management strategies, is the goal of modern stock assessment. The most reliable stock assessments available should underlie fisheries management decisions on the resources and hence the returns from the resource through utilisation (see Sections 2.2, 2.3 and 2.4).

(iii) Responsible fishing should not allow more of the resource to be harvested on average than can be replaced by net growth in the stock. This does not mean that annual catches may never exceed net annual production and, under most harvesting strategies, natural variability and uncertainties are such that catches are likely to exceed production in some years. However, such action should not result in the biomass of the stock or stocks falling below the pre-determined limit reference point (see Section 3.1) at which the risk of resource collapse becomes unacceptably high. Failure to follow this rule will mean that the resource will be depleted with time, leading to lower than optimal average yields and economic returns. If steps are not taken to correct such a situation, the risk of biological collapse, of uncertain duration, and economic waste or extinction of the fishery will be increased to an unacceptable level (*7.2.1*).

(iv) Fish species occur in populations which may be made up of a number of largely self-sustaining stocks, which are effectively genetically isolated from each other by behavioural, oceanographic or topographical features. As far as possible,

UNIVERSIDAD DE SEVILLA
Fac. G.ª e Historia-Biblioteca

fisheries management must address each stock separately and strive to fish each stock sustainably (*7.3.1*) or set an overall exploitation rate that does not drive the components of a multi-species resource to dangerously low levels. Failure to adhere to this will result in a risk that there will be extinctions or severe depletions of individual stocks, even while the population as a whole is in an apparently healthy state. The effective genetic isolation of stocks could mean that such local extinctions are irreversible, and permanent damage is done to the status and productivity of the population as a whole as well as to local fishing grounds.

(v) In addition to avoiding over-fishing of specific stocks, the fisheries management authority needs to avoid actions that will adversely affect the genetic diversity of a stock or population. Long-term fishing pressure above sustainable levels on selected portions of a stock, such as large individuals, can have the effect of reducing the frequency of the selected genetic features, hence reducing the heterozygosity of the stock or population. Such sustained selective effects should be avoided.

(vi) While many fisheries, and many stock assessment and management strategies, focus on single-species or stocks, in reality all species of aquatic resources function within, and are dependent on, communities of differing complexity in terms of the number of species. Therefore, harvesting any one species is almost certain to impact others, either through technological interactions such as the incidental capture of other species during fishing, or through food chain effects such as reducing the abundance of a predator, prey or competitor of other species through fishing. The impact on ecological linkages (e.g. through the trophic chain) between species, may lead to changes in species dominance and affect the dynamic equilibria of the resource system, potentially affecting future options. These multi-species effects need to be considered in responsible fishing, which should aim to ensure that no species, whether targeted, by-catch or indirectly affected by fishing, is reduced to below sustainable levels by fishing (*7.6.9*).

(vii) An important consequence of multi-species effects in fishing is that it is impossible to harvest simultaneously the maximum sustainable yield from each species in an assemblage in a given area comprising a mixture of predators and prey species. Each element of the assemblage has its own biological parameters and characteristics and would require a specific fishing regime catching only that species, a condition impossible to fulfil in practice. In addition, modifying the abundance of predators (or prey species) will affect the abundance of the other elements of the assemblage with which they interact in a way which remains generally difficult to forecast. As a consequence, the optimal multi-species yield from an area will always be lower than the sum of the individual species potential yield. Responsible fishing should therefore not aim to obtain the maximum sustainable yield from each component of a multi-species community as this will lead to over-exploitation of at least some of the component stocks (*7.2.3*).

1.3.2 Environmental constraints

(i) The life stages of fish are affected by environmental conditions, which can influence growth, reproductive and mortality rates. The earlier life stages are particularly susceptible to such influences which can lead to high variability in resource abundance and production on a variety of time-scales. Of most relevance to fisheries management are fluctuations in recruitment from year to year and shifts in ecological regime in which the functional characteristics of the ecosystem, including the composition, abundance and location of the fish community, may change dramatically over periods of decades, driven by environmental forces. Fisheries management must be aware of this variability and, particularly for inter-annual variability, attempt to address it in management plans (*7.2.3*). This requires that fisheries must be able to cope, without adding excessive adverse impact on the stock, with years in which the stock and its productivity are driven to below-average levels by natural environmental fluctuations. Thus, the capacity of the fisheries (i.e. the potential effort which they can exert) must not be determined by and set on the basis of yields which can be obtained in average to good years but on the long-term average, with flexibility to allow for effort reductions in bad years. Failure to address this principle in the case of variable resources will lead to continual pressure to over-exploit the fisheries in below average years and to an overall poor economic performance of the fisheries.

(ii) Environmental variability can also influence the availability of fish to the fishery by, for example, dispersing fish more widely so that they are less available or concentrating them in areas where they are more easily caught. Care must be taken that such changes in availability are not interpreted as changes in stock size, which can lead to incorrect management decisions being made and to excessive and unsustainable catches.

(iii) In most unperturbed ecosystems, an unfished stock will tend to fluctuate around an average maximum level, corresponding to the average carrying capacity of the habitat for that stock. The long-term productivity of a stock is related to this carrying capacity. However, the carrying capacity not only can change through time due to natural variability but also decline as a result of human activity such as destructive fishing methods (e.g. the use of dynamite or cyanide), coastal habitat degradation (e.g. through urban development or destructive trawling in sensitive habitats), modification of river flows, or pollution. This may adversely affect the productivity of stocks, contributing to the risk of over-exploitation. Fisheries management should assess the impact of such influences on the status of fish stocks and their natural habitats and take steps to correct the situation by (i) acting to stop any adverse environmental impacts resulting from human activity (*7.2.2f)* and *g)*; *7.2.3*), (ii) adjusting the fishing pressure to take into account changes in productivity and, if necessary, (iii) restore stocks and habitats to a productive condition (*7.6.10*).

(iv) Alternatively, habitat enhancement can positively influence the productivity of fish stocks through, for example, ecologically sound provision of artificial reefs, appropriate levels of fertilization of lakes, control of predators, restoration of destroyed or damaged coastal, shore-line or river-bank habitat areas, or through improvements in water quality. Attention should be given to maintaining, or restoring as necessary, nursery habitats and migratory pathways, including important longshore and offshore/onshore pathways.

(v) Inland fisheries are particularly influenced by external environmental factors, and responsible use of inland fisheries requires the identification of the primary external factors influencing these fisheries and their impacts on the fish stock or stocks. Such knowledge is required to ensure that appropriate management action can be taken when changes are brought about by one or more of these factors. Such action may include appropriate adjustment of fishing mortality but may also include corrective or rehabilitative action. The most common external factors influencing inland fisheries production include:

- water quantity, both the absolute (e.g. mean) quantity and the distribution of the quantity over time (e.g. seasonal, longer-term cycles, artificial regulation), and

- water quality, which will most commonly change through pollution by toxic chemicals, excessive sediment load or eutrophication.

1.3.3 Biodiversity and ecological considerations

(i) Fishing activities are normally deliberately targeted at one or more species in an ecosystem. However, they frequently also affect other components of the ecosystem through, for example, by-catch of other species, physical damage to the ecosystem or through food chain effects. Responsible fisheries management should consider the impact of fisheries on the ecosystem as a whole, including its biodiversity, and should strive for sustainable use of whole ecosystems and biological communities (*7.2.2d)*).

(ii) As mentioned in 1.1, many fish stocks around the world are already over-exploited and depleted. This results in lower yields being obtained from the stocks than would otherwise be possible (or lowered quality and economic value of the landings), and increases the risk of stock collapse and negative ecological changes. Fisheries management is required by the 1982 Law of the Sea Convention (and by the Code of Conduct) to restore depleted populations (*7.6.10*) to levels above those at which maximum productivity occurs (e.g. to biomass levels higher than the level corresponding to the Maximum Sustainable Yield, MSY). This reflects current thinking that providing a margin for safety that takes into account normal variability and uncertainty requires using MSY as a limit for fisheries management, rather than as a target.

1.4 Technological Considerations

(i) In managing an unstocked, capture fishery, the only mechanism available to maintain the fish stock at a desirable level and with an age structure adequate to avoid recruitment over-fishing, is the control of fishing mortality, i.e. the proportions of the various age groups of the stock which are removed in a given period, usually defined as one year, by the fishery. By regulating the amount of fishing effort exerted, either through catch or effort controls, by regulating the gear or fishing method used and by regulating when and where fishing is permitted, the management authority can control the mass and sizes or ages of fish caught, within the limits of natural variability and uncertainty. This is discussed further in 3.1.

(ii) It is widely recognized that the presence of excess capacity in a fishery increases the pressure on the fisheries management authority to exceed the optimal fishing mortality on a stock, and makes it more difficult to enforce regulations restricting effort. This occurs through social and political pressures to make full use of this excess catching and processing capacity and to retain those people associated with this excess capacity in employment. However, this will clearly only be a short-term solution leading to even greater problems in the long-term. Excess capacity also inevitably involves economic inefficiency. It is therefore in the interests of the users and the resource to maintain potential fishing capacity at a level commensurate with the long-term stock productivity. Mechanisms to do this are discussed under 3.1 (*7.1.8*; *7.2.2a*); *7.6.1*; *7.6.3*).

(iii) Fisheries management authorities must recognise that fishers continually strive to improve the technology they use as well as their cost-effectiveness. This, however, tends to improve their efficiency in fishing. In a fishery regulated by effort control, this can mean that actual effort and hence fishing mortality is, in reality, slowly increasing as fishers discover new ways to become more efficient even though nominal effort (e.g. number of fishing days) is controlled. This phenomenon must be catered for both in control of fishing, where effort control is used as a management tool, and in interpreting effort statistics for stock assessment purposes. Technological progress implies that, in order to maintain fishing capacity and mortality at desired levels, continuous adjustments to meet allowed effort levels may be needed.

1.5 Social and Economic Dimensions

(i) From one perspective, human involvement in fisheries can be seen as having an overwhelming and possibly irreversible impact on the resource. Alternatively, fisheries resources can be viewed as capital stock that if managed responsibly can generate considerable and sustained social and economic benefits. The social and economic dimensions consider the effect of the fishery on people and how to optimize the benefits for the interested parties or interested groups and the society in general. Included in the interest groups are the people who use technology to exploit fisheries resources: to catch and produce fish; to process it in various ways; and to market it or otherwise derive a livelihood from fish. To these interested parties can be added the consumers, lobby groups and other groups which may be indirectly affected by management decisions. Recreational fisheries have considerable social and economic importance in many countries and representative bodies from this sector should be included as interest groups in such cases.

(ii) Responsible fisheries require that the critical factors constituting the social and economic dimensions of the management system be understood (*7.4.5*). The social dimension encompasses a wide range of variables in the human sphere. It is primarily concerned with the interaction between people: how, and why, individuals or groups behave in relation to each other and in relation to the fisheries resource they use or on which they depend. These relations are mediated by a great variety of cultural patterns, habits and customs, instruments of exchange, institutions and individual or group motivations. Further, fisheries are primarily economic activities and the economic dimension encompasses revenues and costs which vary with the level of exploitation and relate to dynamic market forces.

(iii) Social and economic variables interact closely, and any management decision is likely to have effects on, for example, the distribution of income and wealth, the amount and form of employment, the allocation of use rights, the composition and the cohesion of interest groups and sub-groups. More generally, interest group attitudes, positive and negative, towards management regimes will be influenced by management decisions and actions. Fisheries management actions may further affect the contribution of the fishery to critical policy issues such as food security, net foreign exchange earnings, subsidies and other benefits and costs.

(iv) Alternatively, social and economic dimensions can conflict, in which case care must be taken to seek the greatest coincidence between the agreed social and economic objectives of any management plan. Failure to reach a minimum agreed level of compatibility will directly affect the acceptability and implementability of any management plan. The ease with which coincidence is likely to be reached is related to the simplicity of the fishery. For example, coincidence would tend to be more easily reached in pure industrial fisheries, particularly in international fisheries, in which the economic dimension usually predominates. On the contrary, social considerations frequently dominate in small-scale fisheries. Such social considerations can include, for example, the transmission of knowledge, recruitment of crew, investment and credit schemes, solidarity, channels of reciprocal obligations and rights linking individuals of different social status. These factors are in turn often dependent on age, gender, family history, local beliefs and customs. Moreover, in small-scale fisheries, contacts between user groups, political leaders and administrations tend to rely mainly on social processes and institutions. When considering subsistence, small scale and artisanal fisheries, particular attention should therefore be paid to the social conditions, and the specific perceptions, of the participants (*7.2.2c)*). The greatest problems in achieving coincidence in the objectives are likely to occur in mixed (technology and species) fisheries with mixed social, economic and biological objectives.

1.5.1 Social and cultural constraints

(i) Social conditions are subject to constant changes over time and space. Changes can operate on several levels: longer-term cycles of historical change, shorter-term cycles of seasonal change and the immediate month-to-month or day-to-day changes which might be related to weather, employment, demand and supply and other conditions. Such changes will impact the management approach in an interactive manner as people will be affected by the management regime but, in turn, their attitudes to it, and hence its viability, will be influenced by the prevailing state of a range of these dynamic social conditions. Even in traditional societies, where the pace of change may seem slow, these elements in the social system and the historical patterns of resource exploitation must be taken into consideration (*7.6.6*).

(ii) Some social variables are to a degree quantifiable and hence can be measured and subjected to quantitative analysis and modelling. Other variables, however, relate to meaning, values and organization in the social life of the interest groups.

These can be difficult to identify and even more difficult to quantify, often because they are the result of combinations of dynamic elements such as the culture in which people live and its historical development. Examples of such qualitative variables include: motivation of individuals; fishing behaviour, strategies and perceptions of risk; status and political influence in a group or community; perception of moral legitimacy of management practices; and access to information. Knowledge of both qualitative and quantitative variables is the basis for evaluating the compatibility between a management option and the social context in which it has to function (*7.6.7*).

(iii) A first step towards identifying the relevant variables of the social dimension is the identification and selection of those distinct social groups who are concerned with the resource, its use and the benefits which derive from it, i.e the social groups making up the different interested parties. A second step is to analyse how these groups interact and to evaluate how different management interventions may affect each of them. The social groupings will usually vary depending on the type of production unit. In different circumstances, professional skill, kinship ties, age or ethnic group may all play a role in determining the composition of the production unit and the social and economic relations between members of the unit. Such considerations should not be overlooked if the management system is to be accepted.

1.5.2 Economic context and constraints

(i) A paramount objective for the fishery sector as a whole is to realize its full economic potential, as measured over time by the sum of net economic benefits across all producers and consumers, including rent which could be otherwise extracted.

(ii) Under optimum conditions market forces usually ensure economic efficiency. However, optimum conditions generally do not prevail in the fishery sector and there is a need to address carefully the impact, *inter alia*, of externalities and price distortions which might lead to economic overfishing. These are frequently major sources of economic inefficiency, which generally result in rent dissipation and would require management intervention.

(iii) Without proper management mechanisms, fishers tend to have insufficient incentives and information to take into account the effects their activities have on others in the short and long term. This produces a pervasive tendency for over-expansion of fishing effort beyond the point of maximum economic yield. Economic overfishing manifests itself in excessive input allocation to the fishery, inducing, particularly in industrial fisheries, over-capitalization and frequently excess fishing capacity as stocks are progressively depleted. Eventually, in most fisheries a point will be reached where the cost of fishing exceeds the value of the catch. In addition, this typically occurs in a context where fluctuations in fish abundance, market prices and operating costs are continuous and induces cycles of investment and stock depletion. States should therefore strive to prevent or, where necessary, eliminate excess fishing capacity, thereby maintaining fishing effort at levels appropriate to the productivity of the resource or resources (*7.1.8*; *7.6.3*).

(iv) Price distortions may also contribute to over-investment and economic waste and often exacerbate management constraints. Among these are the many subsidies which States provide to the sector in relation to investments or to some key inputs such as fuel and various form of tax exemptions and rebates (see Section 3.2.1).

(v) Externalities are common to the fishery sector. These relate in particular to internal externalities linked to the nature of the stocks and their exploitation. They may be imposed by a user of the resource on another user or group of users, for example, when large efficient vessels fish in the same areas as small-scale fishers, resulting in negative interactions between the two sub-sectors or when mobile gear interacts with fixed bottom gear (*7.6.5*). These externalities may result in substantial shifts in fishing behaviour and strategies of interest groups. They may induce conflicts and costs which can be significant and could lead to lower economic performance.

(vi) The economic performance of fisheries is strongly influenced by the broad economic context. Failure to integrate macroeconomic factors and account for externalities generated from outside the fishery sector is likely to undermine the foundation of fisheries management interventions and again encourage conflicts. Fisheries must be influenced for example by the evolution of exchange rates, trade regulations and changes in fiscal policies. Moreover, and particularly at local level, the fishery sector often competes with other sectors, frequently for resource use but also in relation to other externalities such as the impact on fisheries of environmental degradation by other sectors.

(vii) Conflicts between different users for the use of the same aquatic resources are common (e.g between tourism and fishing in coastal areas or between fishing and agriculture in use of inland waters), and an important task of the relevant authorities is to evaluate current and potential conflicts with a view to minimizing them and obtaining optimal returns from the resource. Inter-sectoral and inter-institutional dialogue is therefore inevitably required and should be sustained, for example, between the fisheries management authority and the finance and planning ministries or within appropriate international fora. Such dialogue and information exchange will enable the fisheries sector to take advantage of, or make appropriate adjustments to, exogenous policy and economic changes. It will facilitate developing coherent and consistent options or proposals to guide the fishery towards participating in achieving the goals determined by macro-economic policies, local development strategies or the evolution of the international context.

(viii) Many existing fisheries management concepts are based on cases where a single State has complete jurisdiction over a fishery. In the case of transboundary fisheries, States tend to act like competing harvesters, each of which is induced to neglect the impact of its own harvesting on the stock and on future productivity. Such activity has the probable consequence of driving the stock and fisheries, on both sides of the boundaries, into the spiral of over-capitalization discussed above. Unless the countries concerned reach a binding agreement or, failing that, nevertheless cooperate (*7.1.5*) to conserve and manage the fishery jointly, the economic performances of each and all participating States are unlikely to be maximized.

(ix) However, poor economic performances, as well as management failures, are frequently directly linked to the complexity of most fisheries (unavoidable by-catches and discards, uncertainty and imperfect information, incomplete and multiple jurisdictions, irreconcilable conflicting objectives). These constraints are often exacerbated by an inability or unwillingness to implement and support the costs of management measures that reduce externalities. Hence, evaluation of economic performances must not oversimplify fluctuations in the economic parameters of the fishery. This is particularly valid for multi-species fisheries in which the various species are likely to interact.

(x) The evaluation of economic performances requires an assessment of all costs and all benefits, direct and indirect (see Section 2.3.3), associated with a fishery or a specific sub-sector, related to the users and to the management authority and generated from inside or outside the sector. Each stage of the management process involves transaction costs such as for gathering information, coordinating participants, resolving conflicts, monitoring conditions and enforcing decisions. The transaction costs of management are likely to vary according to the extent of user participation in the management process (see Section 3.3) and they should also be evaluated.

(xi) In international fisheries, measurement of costs and benefits may include unique considerations. For example, there may be specific costs and benefits associated with the cooperative action required for international management of the fishery. Economic constraints to transboundary management might be, for example, if some non-contracting parties to a cooperative arrangement find it beneficial to avoid management and compliance costs of the agreed management strategy and plan (e.g. costs of carrying observers, of modifying gear and enforcement costs). Such non-compliance could also impose external costs on all contracting parties, such as reduced global catch for the contracting parties leading to displacement of vessels with possible loss of net revenues. Another difficulty with international fisheries is the probable existence of different national interests and objectives possibly leading to conflicts. These differences could include discount rate, production costs, consumer preferences and prices of fish on the national markets.

(xii) Responsible fisheries management requires an assessment of the economic consequences of any management action (*7.6.7*). A primary requirement is to estimate the value of the fishery and incorporate consideration of this value into the various options for the allocation of the resource and the extraction of its rent. The net value to society of a fishery is any returns earned in excess of the opportunity costs of the labour and capital employed. Rent can be obtained by the State or local authority either through taxation (on catch or effort) or by charging users for use of the resource. An alternative might be to leave rents with the industry in such a way that their value could be capitalized into property rights. Whatever the management option retained, the extraction of reasonable rent will usually require limiting usage.

(xiii) In determining the value of a fishery or exploited resource, key issues that need to be considered are the definitions of the management unit and the production units within it (see Section 2.3.3). The definitions must be wide enough to include all economic factors that impact on the fishery. Otherwise, control of one sector (e.g. the commercial sector) may unintentionally transfer benefits to or from another sector (e.g the recreational sector). Further, it needs to be recognized that regulating only a part of the fishing activity on a resource or area, for example the industrial fishery sector, may exert an economic bias on, for example, the artisanal sector, with unexpected consequences preventing achievement of the objectives for the fishery as a whole. Therefore, management units should, as far as possible, consider all interacting fisheries and related activities.

1.6 Institutional Concepts and Functions

(i) Fisheries management institutions, as defined in Section 1.2, must be consistent with the requirements of responsible fisheries management. They should be tailored to the characteristics of particular fisheries and individual nations and, as far as possible, be designed to match the expectations and perceptions of resource users.

(ii) In a broader sense, the institutions may comprise the various sets of relations between individuals or groups of interested parties and the State or States which define their respective rights and responsibilities. These may include rules (e.g. designation of the management regime), mechanisms (e.g. decision-making process) and the organizational support structures that develop and implement the rules affecting the use of the fishery resources. Such support structures may include, for example, a fishery administration, intergovernmental management body, gathering of village elders or committee of users.

(iii) In this document, fisheries management arrangements or organizations are addressed principally through two main categories of institutions: the fisheries management authority and the interested parties. The management authority will be the legal entity which has been designated by a State or States to make the decisions on how the fishery must be carried out and to implement the decisions. It is taken to be accountable for ancillary services, such as resource allocation, consultation with interested parties or determination of the conditions of access to the fishery. The term interested party (or interest group) will generally refer to any party which has been accepted by the State or States or by the management authority on behalf of the State or States, as having a legitimate interest in the fisheries resources being managed.

(iv) In the case of international fisheries, representatives of the States concerned are likely to be the interested parties and must take responsibility for the interests of their citizens, some of who may themselves be members of interest groups within or between States, such as fisher organizations or non-governmental organizations (*7.1.6*). In the case of fisheries within national jurisdiction, States must decide within the fishery and its direct and indirect segments who should be among the interested parties *(7.1.2)*. In both cases, States should recognize that representatives of States or groups which can show a legitimate interest and commitment to long-term management objectives can also be admitted as interested parties *(7.1.4)*.

1.6.1 Institutional context and characteristics

(i) Management institutions may differ widely in nature and serve a wide range of functions, including, as is often the case with traditional institutions, functions besides the management of fisheries. Different institutions will differ in terms of their rules, mechanisms and structures and the manner in which they are assembled. The effectiveness of a fisheries management institution is highly dependent upon the appropriateness of its separate components and the way in which these components interact. The cause of ineffective management often lies with institutional deficiencies expressed both in terms of functions and assemblage. In addition, the perception of the legitimacy of the institutions will affect the extent to which those institutions will be efficient in carrying out their management responsibilities.

(ii) The groups of interested parties should preferably have conditions for membership satisfying specific standards in fisheries management (e.g. production, market, resource conservation, environmental protection, etc.). When based on clearly defined interests or common factors, such as geographical areas, ports, community, gear or resource type, these groups may be more effective and ultimately result in better management performances than if they are poorly defined and heterogeneous in composition. In the fisheries management process, these institutions, or the decision-making mechanisms associated with them, will be the principal point of contact between the individuals constituting the groups of interested parties and the fisheries management authority responsible for developing plans.

(iii) Final responsibility for decision-making usually lies with the relevant political authority. However, responsible fisheries management requires that institutional partnerships enabling various types of collaboration with the interested parties be recognized as possible alternatives to locating the entire set of management responsibilities with pure government structures or arrangements (*7.1.2*).

(iv) Institutional designs for ensuring the collaboration of interested parties in the fisheries management process may range from: setting informational mechanisms aimed essentially at presenting the outputs of the management planning process; through establishing consultative mechanisms designed mainly for the responsible government structure to gather information and receive guidance from interested parties; to establishing specifics sets of mechanisms for sharing management responsibilities with various degrees of delegation of competencies (see Section 3.3).

(v) When feasible, mechanisms should be considered to ensure that the interested parties provide for the cost of operation of the fisheries management authority at a level adequate for it to discharge its functions.

(vi) Institutions are dynamic by nature and need to be permanently monitored, evaluated and adjusted as necessary to ensure ongoing effectiveness and legitimacy. Institutional adjustment is a complex and sometimes uncertain endeavour. For example, the adoption of institutional arrangements that facilitate economic efficiency may encounter obstacles such as cultural values, inertia within producer organizations or political sensitivity to the consequences of change. To tailor management institutions to the state and nature of particular fisheries, care should be taken to adopt approaches that are flexible and which explicitly allow for regular re-negotiation of management measures. States should also review the performance of their management authorities and those international ones of which they are members at intervals normally of every three to five years.

1.6.2 Role and function of fisheries management institutions

(i) A primary function of a fisheries management institution is typically to identify and implement rules and procedures whereby the fishery can be carried out in a sustainable fashion to meet established objectives. Usually these rules mainly translate policy objectives into rights and obligations, supported as necessary by policy instruments. For example, they may determine the nature and extent of users' entitlements or define the conditions of access to a fishery and may be supported by a fiscal scheme to extract rent. Whatever their nature (e.g. formal or informal), identified rules should be embodied by the State or States in the legal regime in force (see Section 4.3.1) or sanctioned at policy level.

(ii) Responsible fisheries management requires the existence of management institutions among which would be one or more explicit fisheries management authority. In particular, functions of any management authority at a minimum should include the mandate for:

- identifying the interested parties and overseeing the formulation of the management objectives;

- translating, in cooperation with the interested parties, these objectives into management plans and defining the criteria upon which decisions and regulatory measures will be based, evaluated and adjusted as necessary;

- ensuring implementation of the management measures through monitoring control and surveillance;

- coordinating the collection and analysis of information and data necessary to allow responsible fisheries management (see Sections 2.2 to 2.4); and

- liaising and negotiating on behalf of the fisheries interests with users of other resources or areas having an impact on fisheries.

The area of competence and the fish resources, fisheries and geographical areas for which the fisheries management authority is responsible and accountable must be precisely defined.

(iii) A fisheries management authority may incorporate sub-structures or subsidiary bodies as required to perform various functions, or it may be constituted of several separate autonomous but interrelated bodies. The presence and number of subsidiary bodies and the nature of the linkages between them will vary substantially depending on factors such as the political, geographical or fishery context and the nature of the authority or mandate assigned. However, when relevant, levels of responsibility among subsidiary bodies should be clearly specified. Regardless of the structure chosen, care must be taken to ensure that efficient channels of communication, interaction and feedback exist between the various components of the management authority, with the interested parties and with any other institutions indirectly concerned with the fishery.

(iv) When States devolve all or part of management functions to local government or groupings, such as management committees, producers' organizations or fishing communities, this delegation of authority must also include clarification of their respective functions and, where relevant, delimitation of the geographical area or management unit falling under each jurisdiction. For fisheries managed at local level, care should be particularly taken to establish unequivocal arrangements on the nature and allocation of access rights, the consultation process, the mechanisms for the collection and analysis of data and information and the structure for compliance and enforcement (see Section 3.3).

(v) General institutional frameworks for transboundary stocks and high-sea fisheries are set out in the 1982 UN Convention on the Law of the Sea and, in the case of high-sea fisheries, are further detailed in the 1995 UN Convention on Straddling Fish Stocks and Highly Migratory Stocks. They are also suggested in Agenda 21 adopted by the 1992 UN Conference on Environment and Development (UNCED). More generally, the nature of responsibilities, mode of operation and structure of international or regional fisheries management arrangements or institutions dealing with marine or inland fisheries should not differ substantially from national fisheries management institutions. The primary differences usually relate to the international nature of the arrangement or institution, frequently resulting in additional complexity. In particular, the States or management authorities concerned must reach a balance between national and common interests and must overcome the generally limited enforcement capacity of intergovernmental institutions.

(vi) If the area of jurisdiction of the management authority, as constituted, only covers part of the stock area or, in such cases where the area of responsibility of a fisheries management authority overlaps with the areas of responsibility of one or more other fisheries management authorities, mechanisms for cooperation or specific

UNIVERSIDAD DE SEVILLA
Fac. G.ª e Historia-Biblioteca

bilateral, sub-regional or regional institutional arrangements or organizations should be set up (*7.1.3*). An integral part of such cooperation must include the collection of relevant biological, social, economic and environmental data in a uniform and accurate way throughout the range of the stock and the sharing of these data (*7.3.4*).

1.7 Time-scales in the Fisheries Management Process

It is important to recognize that fisheries management requires goals, research and actions on a variety of time and political scales (from days to years and from local to intergovernmental). There is considerable overlap between activities at the different scales and frequently the same individuals, groups and institutions will be involved in processes and decisions related to more than one time- and political-scale. These different scales occur within a regional or broader context, in the case of transboundary fisheries, or at a national or local level in the case of stocks confined to a single EEZ or local area.

There are three primary activities, occurring at and involving different scales, which should be explicitly considered by fisheries management authorities.

(i) Fisheries policy and development planning. Fisheries and the optimal use of living aquatic resources are frequently important within national or local economies and also interact with other geographically contiguous social and economic activities or compete for use of common resources such as coastal or riverine habitat, water usage, etc. This macro-policy and macro-economic context requires that fisheries activities should take into account national development planning strategies. It is therefore important that policy and planning decisions are made in full knowledge of the implications, costs, benefits and alternatives for use of the resources. These policy decisions will not include the details of daily fisheries management activities, such as specific control measures, but should provide the broad directions on how the resources are to be utilised and the priorities to be given. The policy or policies would normally include the criteria by which access to resources is granted. For example, the fishery policy could stipulate whether preference in each fishery should be given to small-scale traditional fishers or to large-scale industrial fisheries or to some other arrangement. The development and stipulation of policy is normally the responsibility of government, advised by the management authority and other relevant government departments. Policy should be reviewed regularly (e.g. every 5 years).

(ii) Management plan and strategy. Fisheries policy will normally stipulate the broad directions and priorities to be pursued in utilization of a nation's living aquatic resources. The policy, as it applies to any specific fishery or stock, needs to be translated into a detailed management plan for each fishery (*7.3.3*; see Section 4.1) which includes the stocks being considered, the agreed biological, social, and economic objectives, the control measures and associated regulations, details of monitoring, control and surveillance and other information specifying how the fishery will be managed. The management plan and strategy should be developed by the management authority with full input from the recognized interest groups and should be evaluated and reviewed, including an "audit" of performance, every three to five years.

(iii) Management implementation. The management plan provides details on how the fishery is to be managed and by whom. It should include a management procedure which gives details on how management decisions are to be made according to developments within the fishery, particularly in response to changes in resource status from year to year. For example, the management plan may specify management by a total allowable catch, and the management procedure would then specify how the total allowable catch is to be calculated each year on, for example, the basis of stock assessment using commercial catch and effort statistics and the results of a fisheries-independent survey. Management implementation involves the action and decision-making necessary to ensure that the management plan is put into operation and functions efficiently. It therefore includes responsibilities such as collecting the data necessary to make resource and fishery control decisions, for example determining the annual total allowable catch (TAC) in accordance with the management procedure, licensing of fishers, monitoring, control and surveillance, and liaison with interest groups on the status of the fishery and resources in relation to the management plan. These are discussed further in Section 4.

1.8 Precautionary Approach

The Code of Conduct dedicates *Article 7.5* to the precautionary approach to capture fisheries and species introductions (*FAO Fish.Tech.Pap.*, 350/1, reissued as *FAO Technical Guidelines for Responsible Fisheries*. No.2. Rome, FAO. 1996. 54p.). The concept of the precautionary approach in the context of the protection of the environment was enshrined in Principle 15 of the Rio Declaration of the UN Conference on Environment and Development which states:

> "In order to protect the environment, the precautionary approach shall be widely applied by States according to their capabilities. Where there are threats of serious or irreversible damage, lack of full scientific certainty shall not be used as a reason for postponing cost-effective measures to prevent environmental degradation."

The implications of the precautionary approach to capture fisheries and species introduction have been intensively examined in the Guidelines referred to immediately above. Section 1.6 of these Guidelines summarizes the implications of the precautionary approach as follows.

"1.6 The precautionary approach involves the application of prudent foresight. Taking account of the uncertainties in fisheries systems and the need to take action with incomplete knowledge, it requires *inter alia*:

a. consideration of the needs of future generations and avoidance of changes that are not potentially reversible;

b. prior identification of undesirable outcomes and of measures that will avoid them or correct them promptly;

c. that any necessary corrective measures are initiated without delay, and that they should achieve their purpose promptly, on a time-scale not exceeding two or three decades;

d. that where the likely impact of resource use is uncertain, priority should be given to conserving the productive capacity of the resource;

e. that harvestable and processing capacity should be commensurate with estimated sustainable levels of resource, and that increases in capacity should be further contained when resource productivity is highly uncertain;

f. all fishing activities must have prior management authorization and be subject to periodic review;

g. an established legal and institutional framework for fishery management within which management plans that implement the above points are instituted for each fishery; and

h. appropriate placement of the burden of proof by adhering to the requirements above."

2. MANAGEMENT DATA AND INFORMATION REQUIREMENTS AND USE

(i) The collection of data and information is not an end in itself but is essential for informed decision-making. It is therefore important for the management authority to ensure that the data collected are analysed correctly, disseminated to where they can best be used, and used appropriately in decision-making. Information is also needed to assure the public at large that resources are managed responsibly

(ii) It must be recognized that data and information are required at the three levels, policy formulation, formulation of management plans, and the determination of management actions to implement the policy and plans. These will overlap considerably and each of the three steps will be influenced by what has happened or is happening at the other two levels. Nevertheless, the three processes are distinct, occur on different time scales and require different information to different levels of detail. Where necessary, differences in methods and approaches between, e.g., artisanal and commercial fisheries and at different time scales need to be emphasised.

2.1 General Considerations in the Collection and Provision of Data and Information for Fisheries Management (*7.4.2; 7.4.4*)

2.1.1 Data requirements for different management scales

(i) In general, as the locality and scope of decisions moves from management implementation to management planning to policy formulation, the degree of synthesis and aggregation of information required will increase. At the level of management implementation, details on the current biomass and the age structure and distribution of a stock may be extremely important. However, at the other end of the spectrum, the policy-makers may need to focus mainly on the options for potential annual yield, provided by technical experts, the fisheries' national socio-economic role and their interactions with other macro-economic or macro-policy considerations. These distinctions are discussed in more detail in 2.2 to 2.4.

(ii) The management authority should be responsible for setting-up and overseeing the structures and mechanisms for routine collection and analysis of the necessary data.

2.1.2 Verification or validation of data

(i) Approaches to collecting data for fisheries management vary substantially, depending on, for example, the nature of the fishery, the staff and facilities available, and the social and economic importance of the fishery. Whatever methods are used, the quantity and quality of the data collected will have a direct influence on the quality of the management which can be exercised, and so the most effective use must be made of personnel and facilities available for data collection.

(ii) The verification or validation of data is essential to ensure that it is accurate, complete and gives a true indication of the state or value of the factors under consideration. The problems associated with the collection of fisheries data mean that the risks of collecting erroneous or inappropriate data are very high without careful and statistically valid design and monitoring of sampling approaches.

(iii) Different types of data will need to be verified in different ways. Some examples of methods to validate data include:

- checking logbooks against landings data (e.g. sales notes);
- sampling catches for species composition;
- comparing landings statistics with certificates of origin, trade and commodity production statistics (e.g. processed fish) and similar sources of information;
- inspecting data collection methods by statistical staff;
- interviews with fishers;
- observer schemes;
- reporting from sea on retained catch on entering and leaving the fishing zones;
- developing and implementing the use of vessel monitoring systems such as transponders to monitor the position, catch and activities of vessels; and
- instituting airborne and shipboard surveillance, associated with the boarding of vessels.

(iv) Adequate training and supervision of staff involved in monitoring are essential if the data collected are to be valid. Staff involved in data collection are frequently relatively junior in organizational hierarchies. However, they are also frequently expected to work in remote areas or as the sole observers aboard ships, often with no contact with their supervisors or colleagues for lengthy periods. It is important that they are prepared for this with adequate training and that every effort is made to maintain morale and an awareness of the role of their task within the broader fisheries agency. Regular site visits, incorporating quality control, should be made by supervisory staff to data collection points and regular in-service training sessions held.

2.1.3 Standardization of data collection (*7.4.6*)

(i) Many stocks, and possibly most marine stocks, are not found exclusively within the areas of national jurisdiction of a single State but are distributed across international boundaries. As stated in 1.3.1, stocks must be managed as units or the management actions will almost certainly fail to achieve the desired objectives. Where this requires cooperation between management authorities of different countries, provinces or local agencies, the task of cooperative management is made much easier and more effective if the different partners in the cooperative management all collect data according to common definitions, classifications and methodologies and in a pre-agreed, standardized format, enabling all data to be combined and compared as required.

(ii) Collection of data in a standardized manner will require that the cooperating partners meet periodically to agree on the data requirements, the methods to collect the data, the amount of data to be collected and to review the sample design within each independent jurisdiction. In addition, joint training of staff involved in data collection will almost certainly be advantageous.

2.1.4 Timely distribution

(i) The prompt provision of data in time for appropriate decisions and action to be taken is essential for effective fisheries management. Regular and frequent, typically annual, assessments of fisheries and resources, and a review of appropriate management options in response to changes are essential, and these can only be effective if they incorporate reliable and up-to-date data and information. With due regard for confidentiality requirements (see Section 2.1.5), management authorities should participate in and encourage sharing of information and data amongst different agencies and interest groups with genuine needs for these (*7.4.6; 7.4.7*).

(ii) Collection of appropriate and high quality data can be complex and costly but, in view of the above, fisheries management authorities must ensure, through the provision of adequate support, that the necessary data collection and analysis systems exist and function effectively.

(iii) Particularly where distances between sampling points are great as, for example, with highly migratory or straddling stocks, the potential role of data transmission by radio, fax, *E*mail and satellites or transponders installed on commercial fishing vessels should be considered.

2.1.5 Confidentiality of data (*7.4.7*)

(i) It is usually important to the fishing industry and to individual fishers to know that aspects of the information which they supply to fisheries management authorities are kept confidential, in particular that information or those data which could be used by their commercial competitors to gain an advantage. In view of this, fisheries management authorities must implement policies and strategies which ensure confidentiality of data falling into this category. While aggregated catch data are generally not regarded as being confidential, data relating to the fishing activity of individual vessels or company specific catch rates, fishing localities and fishing strategies are frequently seen as being of potential interest to competitors and therefore need to be kept confidential.

(ii) Fisheries management authorities need to liaise with all providers of data to establish which data should be kept confidential. Failure to do this could result in future problems in obtaining data from companies, falsification of data or similar problems related to or stemming from a lack of trust in the management authority.

(iii) Related to commercial confidentiality, it is usually desirable to ensure that the data collection tasks and structures of a management authority are kept totally separate from the enforcement tasks and structures. Failure to do this will generate a fear amongst fishers that the data they supply to the management authority to facilitate monitoring and assessment of the stocks and fishery will be used against them by the enforcement arm. Again, this could lead to difficulties obtaining data or to falsified or incomplete data.

2.1.6 Costs of collection and collation of data

(i) The collection, collation and dissemination of data should be carried out in the most cost-effective manner possible so as to minimize costs while acquiring the required information. Collection and analysis systems should be based on appropriate statistical designs to ensure that sufficient but only necessary data are collected.

(ii) Duplication in data collection and analysis should be avoided unless deliberately intended for validation purposes or for other reasons related to maintaining quality. Unnecessary duplication is most likely to occur where there are straddling or shared stocks with multiple authorities having common responsibilities.

2.2 Data Requirements and Use in the Formulation of Fisheries Policy

(i) The role of fisheries in the regional, national or local economy must be understood before the best policy decisions can be made. Therefore, there must be a clear understanding of the position and status of fishing in the national economic and social interest.

(ii) Fisheries usually generate benefits in terms of economic return, employment and food production and in terms of recreational opportunities. However, they also generate costs to the community or State which may arise through meeting management requirements, provision of facilities or subsidies, interference with or prevention of other activities in the same area and other causes. Proper policy decisions require up-to-date and accurate information on these factors.

(iii) A summary of the types of data required in the formulation of policy is provided in Table 1.

2.2.1 Types of fisheries, the stocks they depend on and their ecological and environmental context

(i) The structure of fisheries, both within areas of national jurisdiction and beyond, may be complex, targeting on a range of species with different gears, involving different scales of operation and different fishing zones and landing sites. Fisheries may range from a commercial focus to subsistence to recreational. At the level of policy-making, information on the potential magnitude, possibly measured in terms of potential catch, economic value and employment opportunities for each fishery or each stock should be provided. Failure to do this could result in policies which lead to unrealistic social or economic expectations and hence encourage over-exploitation.

(ii) There is frequently interaction between fisheries for different resources in an area, and, as mentioned in 1.3.1, fishery activities on one stock or biological community may impact others. Therefore, fisheries management authorities need to advise policy-makers on the potential implications for other fisheries of any changes in policy for a given fishery (*7.2.3*).

(iii) Fish stocks depend on their environment, and the nature of this dependence may differ for different life-stages. National policy decisions which have implications for the environment of stocks important to fisheries, even if the decisions do not directly involve fisheries, need to be made taking cognizance of their implications for affected fisheries (*7.2.3; 7.3.5*; Fishing operations. *FAO Technical Guidelines for Responsible Fisheries*. No.1. Rome, FAO. 1996. 26p.).

(iv) Lessons can frequently be learnt from the successes and failures of the past, and fisheries management authorities should provide policy-makers with concise histories of the fisheries under consideration, with particular emphasis on problems experienced, previous management strategies followed and their consequences.

(v) The importance of integrated coastal area management (ICAM) and specifically the integration of fisheries into coastal area management is now widely recognized (*Article 10* of the Code of Conduct for Responsible Fisheries; Integration of fisheries into coastal area management. *FAO Technical Guidelines for Responsible Fisheries*. No.3. Rome, FAO. 1996. 17p.). These principles have been developed with an awareness that many coastal areas are subject to open access and frequently to multiple use. The fact that many coastal areas are suffering from increased population pressure exacerbates the problem. Similar problems of multiple use occur in inland waters, necessitating a similar concept of integrated basin management. Fisheries management authorities should be playing an active, and proactive, role in encouraging policy makers to develop integrated policies and plans for the terrestrial and aquatic environments.

2.2.2 Fisheries characteristics

(i) Policy decisions on regional, national or local fisheries should be formulated and made in the full knowledge of the nature of the fisheries under consideration, including the different fishing groups or fleets and their composition, as well as the fishing grounds they use or propose to use.

(ii) Fisheries policy should recognize the potential biological, technical, social and economic interactions between fleets within a specific fishery and between fisheries. Policy should endeavour to minimize negative interactions which could lead to conflict or to poor performance by one or more fishery.

(iii) Policy should take cognizance of the impact of fishing operations on the environment and hence encourage practices which are sustainable and do not result in avoidable damage (*7.2.2f*) and *g*); *7.6.9*, and Fishing operations. *FAO Technical Guidelines for Responsible Fisheries*. No.1. Rome, FAO. 1996. 26p.)

2.2.3 Social and economic information

(i) Humans are an integral part of fisheries systems and fisheries systems cannot be understood unless the social and cultural features and the economic characteristics of the people and communities within the system are understood. Any fisheries management decision is likely to have an impact on peoples' livelihoods and way of life and the purpose of collecting and analyzing social and economic information is to be able to anticipate the nature and extent of these impacts and to make decisions so as to optimize them. Collection and analysis of data on relevant social, economic and institutional factors is therefore essential for responsible fisheries management (*7.4.5*).

(ii) At the level of policy, the decision-makers should have information on the following:

- the interest groups, their features and their interests in the fishery;
- the economic factors related to the fishery, particularly the economic and social dependence of the different interest groups on the fishery;
- details on costs and benefits to the region, State or local area from the fishery;
- the role of the fishery in providing employment to different interest groups;
- the alternative sources of employment and income for the different interest groups or communities;
- the current status of access to or ownership of the resources;
- the institutions currently involved in decision-making within the fishery;
- an outline of the history of the fishery and the historical roles of the different interest groups within that fishery.

(iii) The economic role and performance of fisheries is influenced by the whole regional, national or local economy, and information on these influences is required for wise and responsible policy development. Hence information is required on the main factors contributing to the broader economy, the main factors driving or hindering development within this broader economy and the influences or potential influences of any development on the fisheries sector.

(iv) Social, economic and even institutional characteristics are as dynamic as biological features and tend to change with time. It is therefore important to monitor and provide information on trends in these factors, including on issues such as demographic changes, movements of people, trends in the markets and issues related to costs in order to assist in the development of policies which will not rapidly become obsolete.

(v) Fisheries are frequently marked by conflicts between different sectors or within sectors, and an important role of policy is frequently to determine a fisheries environment in which conflict or the potential for conflict is minimized. Information is therefore required on historical and existing conflicts and their causes, as well as on possible solutions to such conflicts.

2.2.4 Monitoring, control and surveillance (*7.7.3*)

(i) The successful implementation of policy is dependent, *inter alia*, on the effectiveness of monitoring, control and surveillance. Aspects related to this are discussed particularly in Section 4.3.3 and Fishing operations. *FAO Technical Guidelines for Responsible Fisheries*. No.1. Rome, FAO. 1996. 26p. These provide comprehensive information on the requirements for monitoring, control and surveillance of fishing operations, with particular emphasis on larger vessels.

(ii) In setting fisheries policy, the previous records of success and failure in monitoring, control and surveillance in the fisheries of the region, State or local area are important in evaluating the likelihood of success of the approaches proposed in any new policy.

(iii) The costs of monitoring, control and surveillance to a fisheries management authority can be substantial and need to be considered in setting policy. In some cases, the value of the fishery to the users or society may not be sufficient to justify the costs of a proposed monitoring, control and surveillance system and cheaper alternatives may need to be developed. Only approaches which are feasible with the personnel and facilities available to an agency should be considered. Implementing a management plan which cannot be enforced will damage the credibility of an authority, with repercussions possibly spreading to other fisheries.

(iv) Generally fishers and other interest groups will only support legislation and regulations which they consider legitimate. In setting policy, it is important to consult with all recognized interest groups, and to secure their cooperation and participation, to ensure such legitimacy. In addition, the management system should attempt, where possible, to provide appropriate incentives for compliance as well as penalties for non-compliance.

2.3 Data Requirements and Use in the Formulation of Management Plans

(i) The formulation of a fisheries management plan is discussed in detail in Section 4.1. A fisheries management plan is an explicit arrangement between a fishery management authority and the recognized interested parties. It should identify these parties and clarify their respective roles, rights and responsibilities. It should list the objectives agreed on for the fishery and the harvesting strategy, rules and regulations applied to realize those objectives. It should also describe the mechanisms for on-going consultations, the arrangements to ensure compliance and any other information relevant to the management of the fishery.

(ii) The formulation of a management plan should include iterative consultation between the management authority or authorities and the user groups. Stock assessment and modelling approaches, where feasible and appropriate, should be used to investigate the biological, social and economic implications of different harvesting strategies and management options, and the results used to assist in the selection of the most appropriate plan. A summary of data and information essential or desirable in the formulation of management plans is provided in Table 2.

2.3.1 The target stock, or stocks, and its environment

(i) The potential yield from a stock or community is dependent not only on the biological characteristics of the stock and on the environment, but also on the harvesting strategy used, in particular the age structure and species and sex composition of catches, and the timing of fishing in relation to maturity and spawning. These characteristics of the catch can also affect the social and economic benefits derived from the fishery. For example, smaller animals may command a higher price per unit mass than larger animals in some markets while in others, the opposite situation may apply.

(ii) In the formulation of management plans, the management authority, with the participation of the interest groups, should investigate and provide information on the biological, social and economic implications of different harvesting strategies and management options. For biological assessments, data will be required on historical catch and effort in the fishery, the size composition of the catch (translated into age composition, if possible) and the sex and sexual maturity characteristics of the catches.

(iii) While it is sometimes possible to undertake reliable stock assessments using fisheries information alone, independent estimates, or indices, of stock biomass with time generally provide very useful supplementary data on stocks. Where the value of the fisheries being managed justifies it, management authorities should attempt to collect, annually or biennially, fishery-independent estimates of stock biomass or abundance.

(iv) Stocks and ecological communities are influenced by other stocks and ecological communities with which they interact, and changes in their population structure induced by fishing can, in turn, influence these interacting stocks or communities. Where possible, information should be collected, even if only qualitative, on the nature and strength of such relationships to allow the implications of different management plans on non-target species or communities to be evaluated.

Basic data for this would normally include information on trophic interactions through studies on diets of interacting species and on the relative abundances of interacting species.

(v) Information on environments critical in the life history of the stocks or communities should be considered in the development of management plans, particularly for inland waters' species or for marine species where one or more life stage occurs inshore. This will enable consideration, and inclusion, of the possible impacts of other uses of these environments or habitats in the management plan.

2.3.2 Fishery characteristics

(i) Generally, management plans are developed on fisheries which already exist and may have existed for many decades. The fishery on any given stock may be simple, consisting of a single, relatively homogenous fleet, or may be complex, consisting of several different fleet types ranging from, for example, sophisticated factory ships to fleets of artisanal vessels, each fleet using distinct gear with distinct selectivity patterns or fishing different fishing grounds. A management plan needs to consider each of these fleets in terms of their impact on the resources and, in turn, the impact of a management plan on them.

(ii) This requires that data and information be collected and analysed on each fleet such as: the number of vessels or units; their gear characteristics and the selectivity of the gear; any seasonality in fishing; the locality of fishing in relation to the distribution of the stock and other fleets; any navigational or technological aids which assist in fishing; and related factors.

(iii) Systems for sampling landings need to be designed to ensure that the weight of landings and the biological characteristics of the catch, as well as effort, are accurately determined for each fleet. Where there is reason to suspect that discarding of unwanted portions of the catch occurs before landings are recorded, the quantity, species composition and biological characteristics of the discarded portion should be estimated. Observers during fishing operations, or simulated commercial fishing with chartered commercial vessels, are generally the most reliable means of obtaining these estimates.

2.3.3 Social and economic information

This section should be read in conjunction with the information provided in Section 2.2.3, as there is considerable overlap in the data requirements at the level of policy and of the formulation of the management plan. The details provided in these sections should not be taken to be encouraging major involvement of the management authority in the commercial operations of the various interest groups. In most cases, these interest groups will be in possession of the best information on the economic details and trends necessary for successful operation of their fishery interests. Under such circumstances, the role of the fishery management authority will be more to provide advice where necessary and to consider, or facilitate consideration of, issues of common interest between different fisheries interest groups, and between fisheries interest groups and others whose activities interact with the fisheries. This may frequently require capabilities in multi-criteria decision-making and conflict resolution (see Sections 4.1 and 4.2).

(i) Fisheries may also be aggregated into production units, which may not coincide with the interest groups. For example, a production unit may consist of a boat with its associated crew, a single net such as a seine net and those required to operate it or a factory and its management and labour. The impacts of management decisions must be considered in terms of production units as well as in terms of the interest groups. It is therefore essential to identify the types and number of production units in a fishery and to consider the impacts of production units on the fishery and the impacts of management decisions on production units in formulating a management plan. Fishing effort is normally a function of production units and management actions may involve direct action on these units, either altering the number or influencing their mode of operation.

(ii) Interest groups will generally be heterogeneous in structure, and management action may have a different impact on one sub-group within an interest group than on another. For example, the role of women in an artisanal fishery will frequently be different from that of the men. Children may also have a distinct function. Decision-makers, particularly at community level, may be dominated by older men. Information on these differences needs to be collected and analysed to enable evaluation of the impacts of a management plan on all the distinct sub-groups within an interest group. Failure to do this may result in the failure of a management plan because of unexpected social or economic consequences.

(iii) The information necessary to forecast the economic impact of fisheries management actions needs to be collected and analysed for use in the formulation of a management plan. Any action is likely to have different economic implications for the different interest groups, sub-groups and on the fishery as a whole, and these implications need to be estimated and considered. The economic importance of fisheries involves not only people and transactions related to the capture and processing of the fish but also, more generally, the dynamics of investments and markets, and this broader importance must also be considered. For example, in

UNIVERSIDAD DE SEVILLA
Fac. G.ª e Historia-Biblioteca

remote coastal areas whole towns or villages may be ultimately dependent on fisheries and hence affected by fisheries management decisions and plans.

(iv) Economic factors which need to be taken into account for each interest group, and their sub-groups, include:

- benefits, which can include rent earned from sale of the fish at different stages in its capture and processing;

- benefits other than financial, such as opportunities for bartering and factors related to social status;

- costs, including costs such as those associated with harvesting (fuel, repairs, crew remuneration, depreciation of capital, insurance etc.), costs associated with processing (capital depreciation, fuel, electrical power and water, packaging, wages etc.), foreign exchange costs and opportunity costs (the "loss" incurred by not using the money invested in the fishery in some other way, such as simply investing it in an interest-bearing account).

- taxes paid as a result of the fishery and, conversely, subsidies paid into the fishery.

(v) The contribution of the fishery to employment should be quantified. Fisheries management decisions frequently have an impact on formal or informal employment, and the impact of a specific decision, or plan, on this should be considered. Labour in fisheries is frequently seasonal in nature and this should also be explicitly considered. For example, management decisions that ignore the seasonality of fish or labour availability are likely to fail.

(vi) It is important in the formulation of a management plan to have information on the existing institutional structures pertinent to that fishery (*7.4.5*). This requires that information is available on:

- the formal and informal institutions involved in the fisheries and their composition and functions;

- the leadership of different interest groups;

- the decision-making process within the different interest groups; and

- the management mechanisms relating to the fishery at all levels.

Again, this information should be used to assist in ensuring that the opinions and interests of all groups and sub-groups are genuinely considered in the management plan and that the probable impacts on and responses of these groups and sub-groups have been considered and appropriately covered within the plan. Failure to do this will increase the risks of failure of the management plan to achieve its objectives, or result in inappropriate objectives being selected.

(vii) Where there are appropriate institutions in existence, such as traditional structures, these should be used as a part of the fisheries management system. For example, if a community has an accepted and functional system for determining access to a fishery or regulating fishing seasons, this should be incorporated into the fisheries management plan to facilitate both acceptability and implementation. However, care needs to be taken that the institution is appropriate for the function or functions being delegated to it. Institutions which have a customary role may not be appropriate for other non-customary functions and their decisions and actions may not be accepted by the community if they are beyond the normal jurisdiction of the institution. Information is therefore required on the customary roles of existing institutions and on roles which can successfully be transferred to them.

(viii) Finally, social, economic and institutional factors are dynamic and are subject to change. Cultural and political factors may also lead to changes in access distribution or changes in pressures for access. Changes in the market, whether local or international, may result in substantial shifts in fishing behaviour or strategies. All of these may mean that the objectives initially identified for a management plan become rapidly obsolete. Therefore, trends in such important factors need to be described and the information on their implications for a management approach or plan collected and evaluated. Failure to do this may result in a management plan becoming unworkable in a short space of time.

2.3.4 Monitoring, control and surveillance

(i) Monitoring, control and surveillance of fisheries is critical to their successful management, and the widespread failure of fisheries management on a global scale has, in large part, been a result of the inability of regional authorities, States or local authorities to enforce successfully or otherwise ensure compliance with their management regulations and to monitor accurately the behaviour and performance of the fishers. Responsible fishing requires effective monitoring, control and surveillance, which is dependent on the collection, collation and analysis of accurate and relevant data and information.

(ii) Because of the importance of monitoring and control, the implications for it of alternative management plans must be seriously considered by management authorities in selecting the most appropriate plan. Management plans should not be adopted where their implementation cannot be adequately monitored and controlled. Some examples are listed here.

- The use of total allowable catch as a control mechanism requires that all landings must be monitored and catch by species recorded in close to real time, and adequate steps must be taken to prevent discarding of less valuable or unwanted catch or the unregistered transfer of this at sea. This approach requires substantial monitoring and information processing capacity.

- The use of effort control is generally less expensive. It requires, however, accurate fleet registration and close monitoring of fleet performance and of technical or operational developments which could increase fishing efficiency and thus effectively increase effort (e.g. *7.6.2*).

- Use of closed areas or closed seasons requires the capacity to patrol during off seasons or in closed areas to ensure infringements do not occur.

The most appropriate combination of control measures will depend on the nature of the resource, the fishery and the capacity of the management authority.

2.4 Data Requirements and Use in the Determination of Management Actions and Monitoring Performance

(i) An integral part of a management plan should be a management procedure. A management procedure specifies how management actions should be determined and implemented, by describing which data should be collected, how they should be analysed and exactly what management action should be taken according to the results of the analyses. Management actions and the means by which they may be adjusted for changing circumstances, such as biomass fluctuations, should be determined by pre-negotiated management procedures. The specific actions can then be implemented by the management authority without the need for additional consultation and negotiation. A management procedure may also include contingency plans, also reviewed in advance with interested parties, enabling the management authority to undertake speedy and effective action in cases where the stock falls to levels requiring emergency action. If circumstances change before a review of the management plan is due, to the point where the management plan and procedure need to be revised, the requirements become those of 2.3.

(ii) Except under exceptional circumstances of either very high catch or very low biomass, valid and scientifically defensible estimates of the risk to a stock induced by a particular fishing strategy can only be obtained by looking at projections over the medium to long term, typically ten years or more. Such studies should underlie the development of management plans and their associated management procedures. It is not possible to estimate reliably the impact on a stock or community in the medium to long-term of a single *ad hoc* decision on allowable catch or effort. Therefore, management decisions should be made on the basis of a pre-negotiated management plan and procedure, the impact of which has been tested over a suitably long projection period.

(iii) The data required to implement a management plan, usually based on monitoring a stock or fish community and the fishery, are shown in Table 3.

2.4.1 The target stock, or stocks, and its environment

(i) Invariably, an estimate or index of stock size will be required for the adjustment, typically annually, of those control measures included within a management plan. If the management procedure includes a requirement that there should be an annual or seasonal adjustment to, for example, the TAC, total effort, length of a closed season or other measure to regulate fishing mortality, this adjustment will almost certainly need to be made on the basis of the best estimate of the status of the stock. Hence the management authority must ensure that it collects, collates and analyses the data necessary to determine as precisely and accurately as possible this index or estimate, in time for the decision to be made. Probably the most commonly used index of abundance is a mean commercial CPUE figure for the period under review for the fishery or for a representative sub-section of the fleet. The precision of such an index should be commensurate with the risk involved and the possibility of an over-estimation of stock size should be covered under the precautionary approach.

(ii) The spatial distribution of living aquatic resources is dynamic, changing seasonally and sometimes markedly from year to year. Changes in distribution can cause changes in catchability by the fishery or by survey gear. These could be interpreted as changes in abundance, leading to incorrect decisions on management action being taken. Therefore, CPUE data should not be used alone without some additional information on geographic distribution and trends in stock distribution. The best approaches to this are not well defined, but a relatively simple approach that can be taken to incorporating geographic trends is to stratify the area or areas in which a stock is fished into sub-areas, and to analyse each sub-area separately. This will enable evaluation of the CPUE, or survey index, in a variety of localities and thus increase the probability of picking up changes in CPUE in parts of the range brought about by changes in distribution.

(iii) As with data for the development of a management plan, if the value of the fishery can justify it, a valid fisheries-independent estimate of stock abundance provides extremely useful supplementary information. For fisheries which are highly dependent on recruiting age-classes (such as most short-lived species), a survey directed on pre-recruits may be most useful. Surveys should use standard fishing techniques which must remain constant, or be calibrated to each other, for valid estimates of trends or changes in stock abundance to be made from one survey to the next. Experience has shown that it is frequently difficult to avoid changes in fishing technique, and care must be taken in interpreting data where this is suspected to have occurred.

(iv) The Code of Conduct calls for emergency action in the event of natural phenomena having substantial actual or potential negative impact on living aquatic resources (*7.5.5*). Therefore, at least rudimentary information on the status of the key environmental parameters, such as sea surface temperature, climatic conditions (such as wind strength and direction, rainfall, river outflows, etc.) should be routinely collected and analysed to assist in the detection of abnormal phenomena and their influence on the stock which may require particular management measures. Other factors which could be considered include, for example, the water level or seasonal flow patterns in inland waters, changing chlorophyll abundance and distribution, unusual seasonality, oxygen concentrations in low oxygen areas, and the status of key predators and prey on the stocks. Remote sensing has a potentially important role to play in this area.

2.4.2 Fishery characteristics

(i) The nature of the fishery, and the fleets comprising it, here considered to be not only discrete groups of vessels but also of land-based fishers (see Definitions), should have been considered in the development of the management plan. In implementation of the management plan, as with the resources, the role of information on fishery characteristics would be limited to those types of data required by the management procedure to support control decisions.

(ii) The most common use of fishery-related data in implementation of a management procedure is in the use of effort statistics to facilitate estimates of catch-per-unit-effort (CPUE). Again, the method of collection and analysis of catch and effort data should be carefully specified in the management plan. This would normally involve obtaining estimates of the total catch per fleet comprising the fishery and of the total fishing effort, in appropriate units, exerted by each fleet. These would then be interpreted collectively to provide an index of stock abundance, taking note of the considerations discussed under 2.4.1. It may also be possible to use representative sub-sets of a specified fleet component for these calculations where data for the whole fleet are not available. Care must also be taken in this connection in deciding on vessel categories by size, gear type and fishing behaviour. In special cases, for example where one or more fleets are fishing on the same age component of the stock in the same area and with gears of equivalent selectivity, it may be possible to use the data from only one of the fleets for assessment of an abundance index.

(iii) The characteristics and behaviour of the fleet should be monitored to facilitate correct interpretation of CPUE trends. Any changes in fishing grounds, seasonal distribution of effort, gear type or other factors which could influence efficiency of the fishery need to be considered in interpretation of the catch and effort data.

2.4.3 Social and economic information

(i) In general, the social and economic features of the fishery should have been considered early in the development of the management plan, and the agreed needs and preferences of the various interest groups should have been incorporated into it. Hence, implementation of the pre-negotiated management plan should be in the best interests of the different interest groups, including important sub-groups within the groups. Therefore, social and economic considerations would not normally be a major input into the management procedure used to determine annual or seasonal control measures.

(ii) Nevertheless, social and economic forces and dynamics can influence the fishery, *inter alia* giving rise to changes in the behaviour of the fishers (see Section 2.3.3). For example, changes in market preference for size could result in changes in the fishing strategy of the fishery and resulting in changes in CPUE which are independent of stock abundance. On-going liaison with the different interest groups, and monitoring of appropriate social and economic indicators to detect such changes, should be carried out. Where necessary such changes will need to be reflected in the calculations or implementation of the management procedure.

(iii) The management plan should include explicit statement of objectives and these will normally include social and economic objectives. The degree to which the management plan is succeeding in attaining such objectives should be monitored continually and the reasons for and implications of any marked failures in achieving objectives evaluated. Serious failures in performance of a management plan could lead to a decrease in compliance with its provisions or strong pressure to revise the plan without waiting for the scheduled routine review. However, it must be accepted that the natural variability in fish abundance and productivity means that social and economic benefits are likely to be similarly variable from year to year. Therefore, performance of a management plan must be evaluated over a number of years and not abandoned whenever returns fall below the expected average return.

(iv) Notwithstanding the existence of a pre-negotiated and agreed management plan, experience has shown that the interest groups will tend to try and depart from the management plan when it results in decisions that they perceive are not in their immediate best interests, for example if it leads to substantial declines in TAC. In such cases, social and economic arguments are normally used to support a requested departure from the management plan. The management authority should be in possession of the data and information necessary to evaluate these claims of social or economic hardship and, therefore, to be able to weigh them up objectively against negative impacts such as an increased risk to the stock. In general, for the reasons given in 2.4 above, departures from a management plan should only be considered in exceptional circumstances.

2.4.4 Monitoring, control and surveillance

(i) Monitoring involves the collection, measurement and analysis of data and information on fishing activities and is therefore largely covered in the preceding sections. In addition to collecting the data necessary for implementation of a management plan (see Section 2.4), fisheries management authorities must ensure that they are collecting on a regular and continuous basis the data necessary for the next revision of that plan (see Section 2.3).

(ii) Control refers to specifying the terms and conditions under which resources can be harvested and surveillance involves checking and supervising fishing activities to ensure all applicable laws and regulations are being observed by the participants in the fishery.

(iii) Monitoring, control and surveillance needs and approaches will vary substantially from fishery to fishery and even fleet to fleet. For example, the approaches are likely to be very different for small-scale artisanal fisheries fishing in diverse, tropical ecosystems and large-scale industrial fisheries utilizing essentially single stocks.

(iv) Monitoring, control and surveillance require a knowledge of the details of the fishers, the gear they are using, and the port or ports of registry and landing. Such information could, for example, be obtained from a comprehensive frame survey undertaken every two to three years.

(v) Thereafter, at the simplest level, acquiring the data required for monitoring, control and surveillance may involve simply collecting catch and effort information at landing points and encouraging fishers to report any infringements in regulations they observe. However, at the other end of the spectrum, monitoring, control and surveillance may involve the use of dedicated patrol craft and aircraft supported by effective administrative and legal structures. In this case, operations of surface craft and aircraft used for this purpose should be closely coordinated, and patrols should be based on an historical understanding of fishery or fleet operations.

3. MANAGEMENT MEASURES AND APPROACHES

(i) As stated in 1.4, the only mechanism available to maintain the biomass and productivity of a resource at a desirable level, at least in wild capture fisheries, is controlling fishing mortality by regulating the amount of fish caught, when they are caught and the size and age at which they are caught. In regulating fishing mortality there are a number of approaches which can be used, and each one will have different implications and different efficiencies for regulating fishing mortality, impact on fishers, feasibility of monitoring, control and surveillance and other facets of fisheries management. The major options are presented in the following sections.

(ii) In inland fisheries, particularly in rivers and smaller water bodies, it is frequently possible to intensify or enhance fisheries production. This can be achieved by, for example, appropriate and responsible stocking of a water body with supplementary animals of species already present, supplementary fertilization to enhance primary or secondary production without changing ecosystems and biodiversity and elimination of predatory or pest species. These represent steps in a progression from wild, capture fisheries to simple-culture-based fisheries. The principles and approaches to intensification or enhancement of inland fisheries are not discussed further in these Guidelines, but will be addressed in the guidance documentation on fisheries enhancement and aquaculture development.

(iii) In those cases where a stock is fished in areas falling under the jurisdiction of more than one management authority, such as with transboundary or highly migratory stocks, efforts should be made to ensure that management measures are compatible between the different jurisdictions. Failure to observe this may prevent any of the management authorities, or users, from achieving their objectives (*7.3.2*).

(iv) The total amount or mass of fish caught in a fixed period will depend on the concentration of fish in the fishing area, the amount of fishing effort employed during the period and the fishing efficiency of the gear used. This relationship indicates that there are a number of approaches which can be used to regulate the total catch, and hence the fishing mortality imposed on a stock.

- Technical measures, which are restrictions or constraints to regulate the output which can be obtained from a specified amount of effort, for example gear restrictions, closed seasons and closed areas. In terms of the above regulations, these measures generally attempt to influence the efficiency of the fishing gear.

- Input controls directly regulate the amount of effort which can be put into a fishery. In general, inputs are more easily monitored than outputs.

- Output controls directly regulate the catch which can be taken from a fishery and can be seen as an attempt to circumvent the problems associated with defining and enforcing appropriate technical measures and effort regulations by directly limiting the factor of primary concern: the total catch. However, catch controls also have problems, largely associated with monitoring and surveillance.

In most instances, fisheries are regulated by a combination of more than one of the above types of control measures.

(v) An overriding consideration, whichever combination of management measures is used, is the decision as to whether access to the resources will be open or limited (see Section 3.2). The concept presented in 1.4 and 1.5.2 that excess effort and fleet capacity should be avoided in a fishery, coupled with the fact that fisheries resources are generally overfished because fishing capacity is in excess of that required in the long-term, implies that a limitation on total effort with access to a fishery may have to be imposed, while ensuring equity in the process.

3.1 Options to Regulate Fishing

3.1.1 Technical measures

(i) Gear restrictions affect the type, characteristics, and operation of a fishing gear. Some gears have been prohibited outright to (i) avoid increases in fishing capacity through increased efficiency, (ii) avoid some unwanted impact on non-commercial sizes, species or critical habitats, or, very often, (iii) avoid an injection of new technology which could modify significantly the existing distribution of exploitation rights (particularly when these involve new participants). Regulation of gear characteristics such as minimum mesh size or dimensions of mouth opening of nets or traps is generally introduced to control fishing mortality on some particular component of the resource, such as smaller individuals, for example juveniles of the target species or fish of by-catch species. Gear restrictions may also be designed to reduce the total catch by reducing the potential efficiency of the fisher. For example, prohibition on SCUBA gear in some fisheries for sessile bottom dwelling species has this effect. Gear restrictions have an important role to play in making optimal use of a stock or a resource. However, experience has shown that gear restrictions cannot be used alone to ensure sustainability. In addition, impediments to improved efficiency often increase the cost of fishing relative to other fleets and hence may lead to an increased pressure for higher catches to maintain income levels.

(ii) Gear restrictions tend to be species-specific and, for example, a mesh-size designated to capture mature individuals of a smaller species will still catch immature individuals of a co-occurring larger species. The use of subsidiary devices such as by-catch reduction devices (BRDs), turtle excluder devices (TEDs) and grids can be an integral part of responsible fisheries management where, for example, by-catches of over-exploited or threatened species are occurring or fishing is having a negative impact on aquatic communities, and should be utilized by fisheries management authorities as necessary (*7.2.2g*)).

(iii) Area and time restrictions can be used to protect a component of a stock or community such as spawning adults or juvenile stages. As with gear restrictions, they have an important role to play but, unlike gear restrictions, can be used to regulate total fishing mortality on a resource. However, a fisheries management authority would have to monitor available effort and specify appropriate closed areas or seasons such that the effort expended in the open windows did not exceed the sustainable levels for the resource or that restrictions in some time-space windows do not simply lead to transfer of excess levels of effort to other areas in excess of that which was desirable. These measures are subject to the same social and economic problems in open access systems as all other control measures.

(iv) Marine protected areas can have a critical role to play in sustainable fishing. Particularly for territorial species or those with relatively stationary life styles, marine protected areas can be used to preserve spawner biomass above the thresholds (based on biological reference points) necessary to ensure sustained recruitment. Marine protected areas can also play an important role in preserving critical habitats or sensitive life stages of species. Fisheries management authorities need to ensure that the location and extent of marine protected areas are based on clearly stipulated objectives, are appropriate for meeting them and are adequately monitored and controlled.

(v) In addition to their role in conserving the resources, area and time restrictions can be used to reduce or eliminate conflict between different components of the fishery system (e.g. artisanal, industrial, and foreign fleets) or between them and other users. By partitioning fishers or other interest groups into appropriate time or space slots according to the nature of their use or fishing practice, encounters between them can be reduced, thus also reducing the likelihood of conflict (*7.6.5*). Such partition leads, however, to implicit allocations, and conflicts may arise if such allocations are not considered equitable by some users.

(vi) Both gear specifications and area and time restrictions can lead to economic inefficiency and distortions, as they may effectively reduce CPUE below otherwise attainable levels. These measures therefore need to be used as part of an overall strategy developed in consultation with the interest groups. Good scientific information arising from appropriate stock assessment and social and economic studies and projections should be used to guide the choice of technical measures as part of an overall study. Biological reference points, for example in yield-per-recruit analysis, should be considered explicitly where appropriate.

(vii) Minimum size and maturity restrictions can also be used to reduce fishing mortality on life stages of stocks which are considered to require special protection. Where implementation of these regulations (such as on minimum allowable size at landing) requires returning captured individuals to the water, the management authority should determine the survival of returned individuals to ascertain the efficacy of thèse measures.

3.1.2 Input (effort) control

(i) Input controls can include restrictions on the number of fishing units through limiting the number of licenses or permits issued, restrictions on the amount of time units can spend fishing, such as individual effort quotas, and restrictions on the size of vessels and/or gear.

(ii) Placing an appropriate limit on effort and therefore on fishing mortality is seen as being very important in responsible fisheries and has been discussed under Sections 1.4. and 1.5 Some degree of effort limitation by the management authority is a pre-requisite for responsible fisheries, whatever other control measures are in place. Experience has shown that in the absence of a limit on fleet capacity (and mechanisms to stabilize it and compensate for technological progress), the amount of effort expended by industry cannot be effectively controlled. However, where secure and appropriate access rights are in position, the holders of the rights will tend to regulate their inputs (in terms of capacity and effort) to appropriate levels in their own economic interests. Excess capacity is, in general, associated with open access fisheries and tends to diminish once exclusive rights are well established.

(iii) The greatest problems in using input controls alone to regulate fisheries are associated with problems of determining how much effort is actually represented by each fishing unit. Even discrete fleets within a fishery are characterized by considerable variation in the size of vessel (where vessels are involved), the nature of gear and technical and technological aids used, quality of maintenance of vessel and gear, skipper skills and strategies and other factors. These differences make assessing the effective effort exerted in a fishery very difficult.

(iv) In theory, if sufficient data are available, it is possible to determine the relative efficiency of each vessel and fleet by comparing historical catches per unit of effort in a fleet data base. In practice, however, scarcity of data and continual change, often associated with efficiency increases, make such calibrations difficult. This emphasizes the importance to the management authority of the collection of appropriate data on catch and effort (see Sections 2.3.2 and 2.4.2). Undertaking experiments, particularly over short time and space scales, to compare gear efficiency in cooperation with industry can assist in comparisons of effort units.

(v) If the problems of defining effort, determining the amount of effort appropriate to a given resource and monitoring changes in effective effort can be overcome, there are several advantages to this approach compared to regulation by output control, particularly at a coarse or primary level of control. Effort control may also be desirable to avoid the problems of excess capacity, even where output controls are in place.

- Input control is easier and less costly to monitor and enforce than output control, particularly in mixed species fisheries where many output controls (i.e. species-specific quotas) may be required to control fishing effort.

- Associated with the above, mis-reporting of catch is not a serious factor in input control, as there is little or no incentive for the fisher to provide incorrect catch statistics.

- In multi-species fisheries there should be less serious problems of discarding and high-grading as, again, fishers are not regulated on the amount of by-catch landed or reported.

3.1.3 Output (catch) control

(i) Output control is a popular management measure for fisheries, particularly for large-scale fisheries, and there is an even greater interest in extending its application, in association with limited entry, with the current wide-spread interest in individual transferable quotas (ITQs).

(ii) Output or catch control, in theory, allows estimation and implementation of the optimal catch to be taken from a stock by a given harvesting strategy. Given good information on the dynamics of the stock and its response to fishing mortality, the correct catch can, in theory, be estimated to achieve the desired objectives. Catch controls usually involve setting a total allowable catch (TAC) which is then sub-divided into individual quotas by fishing nation (in the case of international fisheries), fleet, fishing company, or fishermen (e.g. in the case of individual quotas).

(iii) In theory, catch control eliminates the need, for control purposes, of estimating the fishing efficiency of all units in the fishery, and of monitoring and responding to changes in fishing efficiency with time, which are features of effort control. However, such assessments will remain necessary, from time to time, to facilitate adjustment of the overall fleet capacity to take into account technological improvements. Without such adjustments, unregulated increases in capacity will increase the incentives for excess fishing and mis-reporting.

(iv) Catch control also has problems in its implementation.

- While catch control may protect the resource, in the absence of limited entry and individual quotas, it does not reduce the social and economic distortions brought about by competing fishers racing to obtain the greatest possible share of the TAC before it is filled.

- Arguably, the greatest problem associated with output control is monitoring the outputs, the catches. The incentive on fishers to mis-report their catches is high when this is the factor used to regulate their own right to fish. Therefore, the management authority has to monitor closely catch per user and in total, to ensure the TAC, and individual quotas where issued, are not exceeded. This leads to the need for a comprehensive, accurate and hence costly monitoring system that collects and analyses data in close to real-time for effective management.

- TACs and individual quotas are normally set and issued for single stocks. In multi-species fisheries this leads to the problem of discards and high-grading, as

TACs and quotas for co-occurring species will be filled at differing rates. If fishers have completed their quota or TAC of a given species but continue fishing for other species, they will be left with little choice but to discard or land illegally catches of the species for which the quota is completed. Again, this leads to the need for an effective, and usually expensive, monitoring, control and surveillance system attached to the management authority to regulate fishing. Arrangements for quota swapping and carry-over of quota or catch from one year to the next may alleviate this problem to some extent.

3.1.4 Some general considerations

(i) From the preceding discussion in 3.1, it should be apparent that different methods of control of fishing have different effects, advantages and disadvantages that will make them more or less suitable under different conditions. There is no single correct approach to controlling fishing, and fisheries management authorities will have to select the option or, more usually, combination of options, which best suits the nature of the fishery and the objectives of the interest groups (*7.6.4*).

(ii) In addition to those considerations which have a direct bearing on the target resource and on the interest groups, fisheries management authorities must be aware of the indirect impacts of fishing and must take steps and select approaches that minimize harmful effects such as waste, discards, catch by lost or abandoned gear, catch of non-target species (by-catch) and other negative impacts on species associated with or dependent on the target species. Particular care must be taken where the affected organisms include endangered species (*7.6.9*).

(iii) Stock recovery is an obligation under the Code (e.g *6.3*; *7.2.1*; *7.2.2e)*; *7.5.5*) once there is information suggesting that stock size is at or approaching levels where reproduction may become seriously threatened. At this point, an explicit recovery plan may be called for, during which the attainment of optimal yields defined by other objectives will have to be secondary in order that the stock recovery period is not prolonged indefinitely. Such a recovery plan requires, of course, definition of limit reference points for the overfished or depleted state, a means for monitoring recovery, a recovery trajectory to aim for and a transition to optimal yield fishing strategy once completed. The advantages of using unusually favourable recruitment to hasten recovery rather than to provide 'windfall yield' should be emphasized.

3.2 Limiting Access

3.2.1 Problems associated with open access

(i) World-wide experiences with fisheries and other free-range resources have shown that open access systems, where anyone who wishes to has a right to exploit the resource, can have severe consequences. In the absence of control, open access systems will invariably lead to over-exploited resources and declining returns for all participants. This has been found to occur in virtually all fisheries under open access, from small-scale artisanal fisheries to large-scale industrial fisheries whether national or international, and has been dubbed the "Tragedy of the Commons".

(ii) Where there is control of overall exploitation by, for example, an overall TAC, or a limitation on total effort by regulating the length of the closed season, the resource may be protected but serious social and economic distortions commonly still arise. Generally, open access systems are characterized by a race to fish in which all participants strive to catch as much of the resource, with or without regulation, as they can, before their competitors do so.

(iii) Under overall regulation, this race to fish leads to features such as shortened fishing seasons, poor product quality and sporadic availability, excess harvesting and processing capacity and increased costs and related negative social and economic effects. Typically, the very high long-term costs of this situation have been borne by society in terms of subsidies, unemployment schemes, rehabilitation of industries following collapses, subsidies to encourage excess fleet deployment abroad, etc.

(iv) The above considerations have been to a large part responsible for the current world status of fisheries consisting of a high proportion of over-exploited stocks and a generally low (and often negative) profitability. Limited access is widely considered to be essential for efficient and responsible fisheries. Associated with various forms of use rights (and property rights), it has become the norm in most systems regulating utilization of terrestrial resources.

3.2.2 Considerations in limiting access

(i) Use rights regimes can be classified into four basic types:

- open access;
- state-regulated access and use;
- communally-regulated access and use;
- private property.

In practice, in most access rights systems within national fisheries jurisdiction the State retains ownership (state property) of the resource, although in inland fisheries there is a considerable degree of private ownership. Where the State does retain ownership, under certain conditions this may involve payments for granting some form of access or exploitation right to users or access to limited numbers of individual fishers, fishing companies, fishing co-operatives, traditional communities or other identifiable user or user-groups. It should not be assumed, when commencing to regulate a previously unregulated fishery, that traditional access rights have not yet been assigned, and determining the evidence of such traditional rights is essential.

(ii) Despite the range of possibilities in implementing systems of access rights, there are some general principles which are related to granting access. There are four primary considerations in considering the nature of access rights for limited access: the nature of the recipients; the initial method of allocation; whether or not the rights should be transferable; and the duration of the allocated right. These are discussed in the following four paragraphs, essentially in terms of the granting of a right of access, but not ownership, by the State or authority delegated with responsibility by the State.

(iii) The State, regional authority or local authority may allocate an access right to a community, individual or company, or a vessel. In general, the allocation of access rights to a community is done to serve social or political goals such as provision of employment or income, or to maintain human populations in remote areas, although there are no reasons why a community may not prove to be as economically efficient as a private enterprise. Allocation of access rights to individuals or companies, if associated with transferability, is likely to generate higher economic efficiency, although this may be associated with a loss of employment opportunities as economic rationalization is undertaken and may displace ownership from coastal communities. One aim in the allocation of fishing rights to vessels can be to maintain employment opportunities since the quota is associated with the vessel; this prevents fleet reduction for the purposes of economic rationalization but would therefore hinder the reduction of over-capacity where this is a problem.

(iv) Where a State or management authority is moving from a system of open access to one of limited access, the greatest problem is almost certainly in determining which of the previous users should be granted access and which denied access. Approaches to this problem can include a lottery which avoids possible problems of favouritism or unfair decisions but makes no allowance for ensuring that the most responsible and effective users are allowed to continue fishing. An alternative approach is to sell or auction the access rights. Where economic efficiency is the primary goal of the fishery and where considerations of equity are not at issue this may be an appropriate approach. However, if the fishery is made up of people from a wide range of economic standings, this approach will clearly favour the most wealthy. Finally, access can be granted on the basis of a selection of specific criteria, including, for example, a proven history of participation in the fishery, performance, e.g. catch above certain minimum criteria, a history of responsible fishing, of social responsibility, etc. In all cases, equity in allocating rights requires that all current fishers be involved in the process. Particular attention should be given to those with long-standing traditions of fishing, especially, where appropriate, to indigenous people and to those local communities highly dependent on fisheries for their livelihoods (*7.6.6*).

(v) The issue of whether or not rights should be transferable is also important and has implications for responsible fishing. The first reaction of governments, when seeking to restrict situations of open access, has frequently been to establish non-transferable rights, but experience has shown problems associated with this approach. In general, transferable rights encourage evolution of the fishery (including a change of actors and a rejuvenation of the sector). They also lead to greater economic efficiency as they allow the more efficient fishers to accumulate greater access through the market. They also provide a mechanism for new entrants into the fishery which, under limited access without transferability, is problematic. The disadvantages of transferability are that it can lead to the formation of monopolies. Where social goals are important, such as the provision of employment to communities, transferability can negate progress towards achieving these goals. As with many other disadvantages associated with access rights, legal provisions to avoid or minimize them may be possible.

(vi) The final consideration presented here is that of the duration of rights to be granted. In general terms, the advantage of granting access rights is that they encourage a sense of ownership in the user, which should lead to a greater sense of long-term responsibility to the resources and fishery, leading to more responsible fishing. This is particularly true if the user can transfer the rights (to a resource to which he or she has contributed to improvements) to his or her descendants or capitalize on such improvements upon retirement. These facets are encouraged by a longer duration of the right so that the user is aware that he or she will gain the benefits for responsible action or pay the price for negative actions with regard, particularly, to the health of the resource. Long-term rights also make it easier for holders to gain financing for their ventures. However, particularly in the absence of transferability, long-term rights prevent the introduction of new entrants and mean that poor initial decisions on allocation may be difficult to reverse. Overall though, assigning long-term rights within a system of limited access is generally the preferred option for responsible fishing.

(vii) Clearly, as with control of fishing, it is possible to develop a system of access rights which best meets the specific objectives, including macro-political and macro-economic objectives, for the fishery in question. The final system for a given fishery should be carefully negotiated with likely applicants and derived, as far as possible, by consensus. However carefully the system is designed, it should be anticipated that there will be disputes, and a method of fair appeal should be catered for. Nevertheless, the ultimate advantages to the users, the resources and the State, of a limited access system rather than an open access system will more than justify the difficulties of striving for the former.

3.3 Management in Partnership

(i) Responsible fisheries management implies attempting to accommodate the interests of a wide range of parties who often represent competing, or even conflicting, interests. It also implies the recognition that the efficiency and implementability of the management measures are often highly dependent on the support gained from the interested parties. In many situations it will be necessary to rely on various collaborative arrangements or mechanisms between States and interested parties as an alternative to locating the entire set of management responsibilities totally within government structures (see Section 1.6). However, ultimate responsibility for decision-making regarding rules and management measures usually remains with the management authority.

(ii) Management in partnership encompasses the various arrangements which formally recognize the sharing of fisheries management responsibility and accountability between a fisheries management authority and institutions either public, such as local level government, or private, such as a group of interested parties. Hence, management in partnership is likely to carry with it a decentralized and unstandardized nature. It often reflects a concern for efficiency and equity at the State or management authority level, coupled with proven capacity for self-governance, self-regulation and active participation at the level of the interested parties concerned. The decision to implement approaches for management in partnership and the extent of self-management delegated to the interested parties should be based on both the characteristics of the fishery concerned and the capacity of the decentralised or local institutions to handle the authority delegated. There may also be a need for the fisheries management authority to provide assistance, including administrative support, to the delegated partners.

UNIVERSIDAD DE SEVILLA
Fac. G.ª e Historia-Biblioteca

(iii) In the context of international fisheries, where the interested parties are primarily the member States of an intergovernmental fisheries management institution, the extent to which the management authority may further devolve or share its decision-making power and responsibilities to other groups of interested parties is likely to be more limited. However, various form of arrangements allowing a certain degree of management in partnership may be desirable. It may be, for example, in the interests of responsible fisheries that industry advisors attend meetings of management authorities (*7.1.6*). Other forms of partnership apply to some arrangements involving a small number of States in the management of a delimited fishery, such as some international lakes or for shared stocks within a bay or a small gulf.

(iv) Arrangements for management in partnership may be particularly valuable in small-scale fisheries, where the management authority is unable to provide cost-effective service. Management in partnership may also be requested by the interest groups when they are paying for the cost of management. Such arrangements encompass a wide range of possible degrees of intervention by the State. On one extreme, such an arrangement may simply formally recognize an existing system of fisheries management at local level, such as a traditional or customary system, and involve no further intervention or support from the State. At the other extreme, the delegation of authority under the partnership arrangement may require full support from the State in the form of financial and logistic support for the process from the formulation to the implementation and monitoring of the management plan. Within this range, various sets of responsibilities, accountabilities and functions could be devolved, including, for example, selecting appropriate management measures, assigning or registering of fishing rights, and the enforcement of local fishing regulations.

(v) Within the fisheries management planning process, the management authority should identify and consider the situations for which management by some form of partnership is likely to be an effective and sustainable policy option. In doing so, care must be taken to evaluate case by case which type of partnership arrangement is likely to yield the desired long-term returns. This decision should be based on criteria such as the potential for:

- greater reliability and accuracy of data and information;
- more suitable and effective regulations;
- enhanced acceptability of and compliance with management measures;

- reduced enforcement costs;

- reduced conflicts; and

- strengthened commitment to and participation by concerned interested parties.

(vi) However, consideration must also be given to the difficulties which are frequently encountered with management in partnership, particularly at the early stage. Attention will need to be given to, for example:

- higher requirements for institutional change or adjustment than those for purely State control;

- probable increase in transaction costs as a result of longer negotiation processes;

- higher risk to the resources if resource users are not properly organized or do not have the necessary capacity;

- probable diminished administrative support for implementation; and

- risks of greater political or lobbying interference at the local level.

(vii) As a complex endeavour, establishing and implementing partnership arrangements should, as for other management processes, follow a structured approach involving research, consultation, decision-making and institutional reform. Approaches should be flexible to fit specific situations, countries, fisheries and fishing communities. They should also allow for gradual implementation, possibly driven by the accumulation of formal knowledge by the responsible interest groups on the relevant social, economic and environmental issues.

(viii) For fisheries to be managed at local level, an effective legal instrument will be required to define clearly the respective role and functions of the local authority and management groups concerned (e.g cooperative, user committee, traditional community). This instrument should also clearly demarcate the physical territory or fisheries management unit over which the management group will exercise its management functions. This territory should, as far as possible, coincide with the existing realm of activity of the group concerned.

(ix) Within the management group, it is also important to ensure that the conditions for membership are explicit and that they reflect, as appropriate, the social and economic cohesion of the group. Whenever possible, the management group should not be so large as to unduly hinder consultation and decision-making.

(x) Mechanisms should also be established to verify from time to time that the social and economic benefits derived from utilizing a partnership scheme equal or exceed the costs of investing in related activities and that those costs are equitably supported by the participants in the management scheme. Such mechanisms could be part of a wider coordinating arrangement with representation by the interest groups concerned and representatives of the management authority to monitor the partnership arrangement, assist in resolving conflict and establish rules.

4. THE MANAGEMENT PROCESS

4.1 Formulating Management Plans to Reflect Selected Objectives and Constraints

(i) A fisheries management plan is a formal or informal arrangement between a fishery management authority and interested parties which identifies the partners in the fishery and their respective roles, details the agreed objectives for the fishery and specifies the management rules and regulations which apply to it and provides other details about the fishery which are relevant to the task of the management authority (see Section 1.6.2). The relationship between the management plan and the fisheries policy and objectives is discussed in Section 1.7.

(ii) It is suggested that management plans reflecting the management objectives should be drawn-up for all fisheries (*7.3.3*). These management plans will then serve as a reference and information source for the management authority and all interest groups, summarizing the current state of knowledge on the resource, its environment and the fishery, and reflecting all the decisions and actions agreed upon during the course of consultations between the management authority and the interest groups. Ensuring plans are developed and implemented for all fisheries helps to avoid planned management measures on one fishery creating unforeseen problems and externalities in a neighbouring fishery for which no plan is available (see Section 1.5.2).

(iii) The considerations which would normally be included in a management plan are indicated in Table 4. The details of the data required for the formulation of a management plan are discussed in Section 2.3.

4.2 Identifying and Agreeing on Objectives for the Fishery

4.2.1 The need for consultation with recognized interest groups

(i) In most fisheries, governments have the primary responsibility and accountability for making decisions in relation to fisheries management. However, such decisions should be preceded by a number of processes, the details of which will vary according to the nature of the fishery.

(ii) It is essential that the decisions made on objectives and management measures for a fishery should reflect the best scientific information available (*7.4.1*). The amount of information will vary according to the nature of the fishery and the capacity of the management authority. However, in all cases, every reasonable effort should be made to have sufficient information to make informed decisions which reflect the probable productivity of the resource and the nature of its environment. Failure to do this will substantially increase the risks of biological, social and economic damage, unless the precautionary approach is invoked. Wherever the information (or the precision of the advice) is deemed insufficient, a precautionary approach should be adopted (see Section 1.8).

(iii) The utilization of living aquatic resources and the management of this utilization should be seen as partnerships between the management authority and the interest groups (see Section 3.3). The objectives should reflect the reasonable desires of the interest groups, within the constraints imposed by the biological and ecological limitations of the resources and the overriding objectives of national planning. Therefore, consultation and joint decision-making are essential in determining the objectives.

(iv) Many reasonable objectives will be mutually incompatible (see Section 1.5). For example, maximizing average yield from a fishery is incompatible with minimizing biological risk to the resource or minimizing impact on other stocks, such as those predators also dependent on the resource. Similarly, maximizing the economic returns from a fishery may be incompatible with maximizing employment opportunities. Therefore, interest groups will frequently have opposing objectives. In order to achieve maximum compliance and cooperation from all parties, it is important to arrive at a compromise which will at least be accepted, if not without reservation, by all or most of the interest groups. This will require open and transparent decision-making (*7.1.9*) and close consultation with all recognized interest groups and the application of effective decision-making procedures and approaches.

(v) The interest groups may include non-fishery groups and, in the case of coastal and inland fisheries, frequently will include such groups. Where a fishery is substantially impacted by external environmental factors, it is important to reflect this and to include interest groups or authorities responsible for the external factors in the management plan. There are four possible scenarios to be considered within the time frame of the management plan:

- further changes in the external factors are likely which will further modify the fishery ecosystem;
- further changes are likely but will not markedly affect the fishery;
- external influences on the fishery will diminish;
- the situation is likely to remain stable.

Different management responses will be required depending on which of these scenarios is believed to be correct. Liaison with the external interest groups is important to: identify the most likely scenario; through discussion to avoid undesirable changes or promote favourable changes; estimate the most desirable management options within the macro-economic context of the region, State or local area; or to negotiate compensation to adversely affected interest groups.

4.2.2 Determining the appropriate management measures

(i) A variety of management measures is available to fisheries managers, and each of them will have different implications for the resource and for the objectives set for the fishery (see Sections 3.1 and 3.2). The most appropriate set of management measures should therefore be selected to facilitate achieving these objectives. This will require careful consideration of the effects and implications of the measures referred to in Sections 3.1 and 3.2.

(ii) In evaluating management measures, it is necessary to consider their implications for the biological, ecological, economic and social objectives of the fishery under consideration (*7.2.1; 7.2.2; 7.2.3*).

(iii) Any fishery exists within a broader ecological, economic and social context (see Sections 1.5 and 2.2). Failure to recognize this and to adopt management measures which are consistent with the policies, objectives and management approaches of the broader geo-political zone could result in failure, or reduced efficiency of the fisheries management strategy, or conflict between different users (*7.2.3*).

(iv) Similarly, management measures need to reflect the macro-economic policy for, for example, the catchment area, river basin, coastal zone, local area or State (see Section 1.5.2).

(v) There is seldom a single correct set of management measures that should be applied to a fishery; rather, there are trade-offs between desirable, less-desirable or undesirable effects. The most appropriate set of management measures is that which maximises the desirable effects and minimizes the undesirable effects for a particular fishery and the objectives set for it. It is therefore necessary, for responsible fisheries, to investigate the costs (in the broader sense) and the benefits of different sets of management measures to identify those most appropriate for the fishery (*7.4.3*; *7.6.7*).

(vi) An important consideration in the implementation of a management plan is that any controls or constraints on the fishers or other interested parties should be implementable by the management authority. This requires that the management measures are feasible and that the management authority and interest groups have the capacity to put them into effect. For example, management by catch quotas will almost certainly fail if the management authority does not have the human and financial resources to monitor catches, or the use of closed areas will not work if the authority is unable to patrol the areas and prevent illegal activities within them. Therefore, the implications for monitoring, control and surveillance of the management measures being considered should be taken into account.

(vii) As with setting objectives, the management measures in operation form a part of the social contract or arrangement between the management authority and the interested parties. Consideration and selection of the set of measures for a given fishery should be undertaken openly and transparently and with full participation by all recognised interest groups. Failure to observe this recommendation could lead to non-compliance by all or some interested parties. The deliberations and final decisions should again be taken on the basis of the best available scientific information, including information on the biological, economic and social implications.

4.2.3 Reviewing the management plan

The status of the resource, the circumstances and priorities of the interest groups and the national circumstances and priorities of any geo-political zone change with time. This means that management objectives and measures can also become obsolete or inappropriate with time. For this reason, a regular evaluation should be undertaken of the effectiveness and efficiency of the management plan, normally every three to five years with revisions made as necessary (*7.6.8*). Such revisions should be undertaken with all the normal pre-requisites for decision-making discussed in this document.

4.3 Implementation

4.3.1 Effective legal and institutional framework (*7.1.1*; *7.7.1*)

(i) Within the context of these Guidelines, the term "legislation" is used in the broadest sense, encompassing all types of national and local laws and regulations. What follows here is intended to provide general guidance only and much will depend on whether the country in question has a civil law, common law, or other legal system, and whether or not it is a federal system. The term "legal regime" covers, as appropriate, these as well as international legal instruments. The relevant provisions of any legal regime which relate to the management of a fishery essentially guarantee that the rules, i.e. the general terms and conditions under which the fishery should be managed and the mechanisms that regulate conflicts, enjoy the force of law. Those provisions are usually framed, revised or amended to reflect the agreed fisheries management policy. The relevant segments of any fisheries management legal regime should not depart from reflecting the desired medium- to long-term management objectives for that fishery.

(ii) From a formal standpoint, at national level the primary legislation (typically a Fisheries Act) is usually approved by the Legislature (e.g. Congress, Parliament). It is generally broad in scope and typically lays down principles and policies (see Section 1.7). It may also reflect varying degrees of detail of implementation, such as the main features of a specific mechanism (e.g. that controlling the allocation of fishing rights). Legislation adopted by the Government through a delegated law-making authority generally sets out the substantive and procedural details of implementation of the provisions of the primary legislation. These are usually referred to as regulations (which may consist of rules, orders, decrees or by-laws).

(iii) Responsible fisheries management requires that the primary fisheries legislation should, as far as possible, not be subject to frequent changes. It should include reference to establishing fishery management plans and some indications of the modalities of the planning process. The primary legislation should therefore define the institutional structure for fisheries management and should empower this structure with the corresponding authority. Hence, in order to implement successfully fisheries management decisions, legal clarity is essential as to who is entitled to administer and control the use of the fisheries resources.

(iv) At national level, the above paragraph implies that the primary fisheries legislation should spell out precisely the functions, powers and responsibilities of government or other institutions involved in fisheries management, including the delimitation of their jurisdiction. Ideally, the legal regime should provide the basis for one or more fisheries management authorities to formulate, monitor and implement fisheries management plans, including the necessary powers to formulate appropriate management measures and enforce the related fisheries regulations. Where more than one management authority exists, overlapping jurisdictions should be avoided as far as possible.

(v) In specifying jurisdiction, it is necessary to define the policy-making entity, the geographical area the policy covers, the interested parties likely to be bound by the policy, the institutions respectively responsible for implementing and for enforcing the management plan, and how inter-institution jurisdictional disputes will be resolved. Failure to do so will inevitably lead to overlap and conflicts both between sectoral management institutions and between different tiers of government, all claiming jurisdiction in respect of common matters. Consideration should also be given, as appropriate, to reducing administrative inefficiencies and duplications. More generally, any assessment of a fisheries management structure should emphasize the relationship between the characteristics of the institutional system and its likely effectiveness.

(vi) Routine management control measures (e.g. closed seasons, size limits, permissible effort) which may need frequent revision should be spelt out in subordinate legislation, such as regulations, which are, or should be, fairly easily changed to reflect changes in fisheries management needs. A similarly flexible approach could be applied to other matters of procedure. In many cases, keeping regulations simple with a clear connection to the relevant management issue and ensuring that procedures for implementation are fair and transparent, are likely to encourage compliance by interested parties.

(vii) The management authority should continually monitor the suitability and cost effectiveness of fisheries regulations and evaluate them in detail when modifying management plans for specific fisheries (*7.6.7*). Those provisions which appear to be obsolete or not enforceable should be amended as required. Further, the laws and regulations implemented should be readily enforceable, and the jurisdictional and administrative procedures supporting enforcement should be fair and transparent (*7.1.9*). Failure to consider properly these issues is likely to undermine the credibility and acceptability of the overall fisheries legal regime and the level of compliance.

(viii) All relevant international legal instruments, and in particular the 1982 UN Convention on the Law of the Sea and the 1995 UN Convention on Straddling Fish Stocks and Highly Migratory Fish Stocks, should be taken into consideration when preparing or amending fisheries management legal regimes. When relevant, and particularly with regard to inland fisheries, full use should be made of the provisions which exist in international instruments or arrangements providing mechanisms for integrated management of fisheries resources at river basin level. The coastal and riverine zones are usually covered by several overlapping maritime, water, forestry or other legal regimes. The fisheries legislation should be in harmony with the general body of legal instruments covering these activities. Particular attention should be given to other legal regimes in the context of integrated coastal area management, either at domestic or international levels.

(ix) Where the management policy includes some form of formal partnership arrangement with one or more interested party, potentially useful principles rooted in traditional practice could provide valuable guidance to national legislators and should not be overlooked (*7.6.6*).

(x) Appropriately widespread consultation should be undertaken with the interested parties during the process of formulating or amending legal provisions relating to fisheries management (see Section 4.2.1). Such collaboration should ideally be accompanied by acceptance by interested parties of their obligations with respect to management measures. Failure by the interested parties or their representatives to abide by the regulations set by the authority must lead to penalties which should be set at a level to be effective deterrents (*7.7.2*). Such deterrents should include, where appropriate, the loss of the right to participate in the fishery.

(xi) In order to encourage compliance it is important for the management authority to ensure that the laws and regulations are distributed to all interested parties and, where necessary, adequate attention given to ensuring they are fully understood (*7.1.10*).

(xii) Another important function of any fisheries legal regime is to establish the institutional arrangements and procedures necessary to reduce potential conflicts and facilitate their resolution when they occur. When disputes do occur, allowance should be made for recourse to formal procedures and a transparent dispute settlement process. These could include hearings and, as appropriate, the setting of compensation. However, in view of the social and economic nature of most conflicts related to fisheries management, attention may frequently need to be given to achieving a balance of interests when resolving conflict, rather than attempting to identify a correct and an incorrect party. When the authority for managing a fishery is devolved to local or community levels, an instrument of agreement should ideally be negotiated prior to the devolution, by which interested parties accept binding dispute settlement, where necessary, by a third party agreed upon between themselves and the management authority.

4.3.2 Effective administrative structure (*7.7.1*)

(i) The preceding sections have indicated that fisheries management requires the capacity to:

- collect, collate and analyse information on the status of the stocks, the nature of catches and landings and the nature of the fishery;
- collect, collate and evaluate information on the economic and social importance and impact of the fishery;
- in conjunction with other relevant authorities, consider the impact of the fishery on the management of the geo-political zone (e.g. coastal, catchment, economic grouping) as a whole, and the impacts of other activities in this zone on fisheries;
- liaise, discuss and make joint-decisions with all groups interested in the fishery;
- facilitate the formulation of policy relating to the fishery;
- coordinate the formulation of management objectives and management measures, taking cognizance of the preceding factors listed above;
- review the objectives and management measures on a regular basis;
- implement the measures, requiring monitoring, control and surveillance of the fishery.

(ii) The above tasks clearly require adequate capacity and facilities, which will require adequate financing. The Code of Conduct suggests that, where appropriate, efforts should be made to recover the costs of fisheries management, including conservation and research, from the interest groups who are deriving benefits (*7.7.4*).

(iii) In determining the scope of the above activities, the management authority should bear in mind the value of the fishery to the State, sub-region or region, and develop management approaches which are consistent with this (see Section 1.5.2).

Where the management authority is dealing with trans-boundary or highly-migratory stocks, the States and regional or sub-regional bodies should determine between them how the activities will be funded.

(iv) In addition to recommendations included under Article 7, the requirements relating to research by the management authority are listed in the Code under Article 12. This article recommends that States should establish a suitable institutional framework to identify applied research requirements (*12.2*) and should ensure that research undertaken meets internationally accepted scientific standards (*12.6*). Article 12 lists the following areas where research is required:

- biology, ecology and environmental science related to the fisheries (*12.1*);
- fisheries technology, including gear selectivity, environmental and biological impacts of gear (*12.1; 12.10; 12.11*);
- aquaculture (*12.1*);
- economics and social science (*12.1; 12.9*);
- nutritional science (*12.1*);
- role of fish and fisheries as sources of human food (*12.7* and *12.8*);
- health aspects of fish as food (*12.8*).

(v) Most of these research activities are an integral part of ensuring that fisheries management decisions are based on the "best scientific information available" (and possible), and these activities are integral to resource monitoring and assessment. Therefore, the development of appropriate research capacity and on-going collection, collation and analysis of fisheries and fishery resource data and research results is critical to the implementation of responsible fisheries management.

(vi) It has widely come to be recognized that effective, sustainable fisheries are only possible if there is close cooperation and mutual acceptance between the interest groups, probably dominated by fisheries interests, and the management authority. It has also become evident that the debate between interest groups is made easier when all such groups have a really significant interest in the matter to be debated i.e. something valuable to lose. The authority is responsible for ensuring that only significantly interested parties are allowed to participate in the consultation and that this consultation takes place and leads, as far as is possible, to consensus and optimal decisions. This will require the establishment of the necessary structures and responsibilities within the management authority to:

- identify the valid interest groups;

- set up discussion and joint decision-making bodies, with clearly defined responsibilities relating to setting objectives and formulation of management plans, with appeal procedures and with formal communication channels, and to ensure that they meet on a regular basis;

- ensure adequate dissemination of research results, fisheries statistics, fishery plans, other rules and regulations and other material important to ensuring all interest groups are fully informed on the fishery and its management, and hence in a position to fulfil their responsibilities;

- publish and disseminate annual reports of the fisheries management authority.

4.3.3 Effective monitoring control and surveillance (*7.7.3*).

(i) The purpose of a monitoring, control and surveillance (MCS) system is to ensure that fishery policy in general and the conservation and management arrangements for a specific fishery are implemented fully and expeditiously (*7.1.7*).

(ii) There is no unique solution to the design and implementation of MCS systems. While they are based on common principles and goals (see e.g. Sections 2.2.4; 2.3.4; 2.4.4), they need to be modified and tailored to the characteristics and needs of each specific fishery, varying substantially, for example, from artisanal to industrial and from concentrated to dispersed. Fisheries that characteristically have very mobile fleets targeting highly migratory species of fish will require sub-regional or regional co-operation in conservation and management and hence also in MCS. Such cooperation should include measures to deter the activity of vessels flying the flag of non members or non participants in established organizations or arrangements from engaging in activities undermining the effectiveness of the agreed management measures (*7.7.5*).

(iii) A range of separate or inter-linked activities of varying degrees of sophistication can be implemented as part of an MCS system. Some of these activities are simple and inexpensive, involving, for example, collection of catch and effort information at landing points, exchange of data for the same fleets if they are operating in adjacent EEZs and encouraging fishers to report infringements. At the other end of the spectrum are costly and sophisticated activities involving dedicated

MCS vessels and support aircraft (see Fishing operations. *FAO Technical Guidelines for Responsible Fisheries*. No.1. Rome, FAO. 1996. 26p.). Transponders, providing cost-effective and immediate information on vessel location and activities, are already being used for MCS in some fisheries and their application is likely to increase rapidly.

(iv) It can be anticipated that the need for surveillance to ensure compliance with regulations will diminish with greater participation in the management process by fishers and other interest groups. With greater shared responsibility, individual fishers and others will tend to infringe regulations less frequently and will also assume greater responsibility for surveillance themselves, reducing the responsibility to be borne by the overall management authority.

(v) MCS in small-scale fisheries presents a range of unique problems which relate to large numbers of widely dispersed fishers operating within a fishery. An important approach to MCS in such fisheries is, where possible, to foster a strong local awareness of and identity with the need for conservation and management. Through community-based rights of access to resources and other communally-agreed management measures, effective cooperative MCS systems can be developed.

(vi) In industrial fisheries, authorized fishers need to be provided with incentives to elicit voluntary compliance with fishery policy and agreed management plans and management measures, in order to reduce the operational requirements and costs of MCS. To achieve this objective, action must be taken, where not already done, to convert fishers from open-access status to legitimate holders of access in a controlled and restricted fishery. In this way a sense of part-ownership of, and hence responsibility for, the resource can be developed in fishers (see Section 3.2).

(vii) Notwithstanding the above, it is unlikely that the need for surveillance, or at least coordination of surveillance, by a management authority will ever disappear completely. A distinction should be made here between the need for consultation with fishers before developing regulations and subsequent enforcement which should be complete, impartial and not subject to 'special intervention'. Thus, in the event that violations do occur, there must be provision for swift, impartial and appropriate enforcement action to deter fishers from further violations of the agreed management and control measures, including, where appropriate, the loss of fishing rights.

(viii) Because fishers are the primary beneficiaries of MCS programmes, the practice in some countries has been that they bear some, if not all, of the costs of MCS. Where a decision is made to implement this approach, it should be introduced incrementally, and the fishing industry required to assume progressively higher shares of MCS costs. Clearly, this type of measure will be most acceptable and effective if fishers are involved in the management process from policy through to decisions on MCS and enforcement strategy.

(ix) It should also be recognized that not only does an effective and well planned MCS system enhance fisheries conservation and management but should also lead to improved safety for vessels and crew and permits the real-time transfer of market information which can be beneficial to the fishing industry as a whole. This consideration has proved to be a significant one in some industrial fisheries in developed States.

DEFINITION OF SOME KEY TERMS

Biological diversity or ***biodiversity*** means the variability among living organisms from all sources including, *inter alia*, terrestrial, marine and other aquatic ecosystems and the ecological complexes of which they are part; this includes diversity within species, between species and of ecosystems. Diversity indices are measures of richness (the number of species in a system); and to some extent, evenness (variances of species' local abundance). They are therefore indifferent to species substitutions which may, however, reflect ecosystem stresses (such as those due to high fishing intensity).

Biological resources include genetic resources, organisms or parts thereof, populations or any other biotic component of ecosystems with actual or potential use of value for humanity.

Discards are those components of a fish stock (see below) thrown back after capture. Normally, most of the discards can be assumed not to survive.

Exploitation rate, applied on a fish stock, is the proportion of the numbers or biomass removed by fishing. If the biomass is 1000 tons and the harvest during a year is 200 tons, the annual exploitation rate is 20%. See also ***fishing mortality***.

Fisheries management authority is the legal entity which has been assigned by a State or States with a mandate to perform certain specified fisheries management functions.

Fisheries management organizations or arrangements are international institutions or treaty arrangements between two or more States that are responsible for fisheries management, including the formulation of the rules that govern fishing activities. The fishery management organization, and its subsidiary bodies, may also be responsible for all ancillary services, such as the collection of information, its analysis, stock assessment, monitoring, control and surveillance (MCS), consultation with interested parties, application and/or determination of the rules of access to the fishery, and resource allocation.

Fishery can refer to the sum of all fishing activities on a given resource, for example a hake fishery or shrimp fishery. It may also refer to the activities of a single type or style of fishing on a particular resource, for example a beach seine fishery or trawl fishery. The term is used in both senses in this document and, where necessary, its particular application is specified.

Fishing capacity is a concept which has not yet been rigorously defined, and there are substantial differences of opinion as to how it should be defined and estimated. However, a working definition is the quantity of fish that can be taken by a fishing unit, for example an individual, community, vessel or fleet, assuming that there is no limitation on the yield from the stock.

Fishing effort represents the amount of fishing gear of a specific type used on the fishing grounds over a given unit of time e.g. hours trawled per day, number of hooks set per day or number of hauls of a beach seine per day.

Fishing mortality is a technical term which refers to the proportion of the fish available being removed by fishing in a small unit of time. Fishing mortality can be translated into a yearly ***exploitation rate*** (see above) expressed as a percentage, using a mathematical formula.

Fish stock or ***fish resource*** means the living resources in the community or population from which catches are taken in a fishery. Use of the term ***fish stock*** usually implies that the particular population is more or less isolated from other stocks of the same species and hence self-sustaining. In a particular fishery, the fish stock may be one or several species of fish but here it is also intended to include commercial invertebrates and plants.

Fleet is used broadly in this document to describe the total number of units of any discrete type of fishing activity utilising a specific resource. For simplicity, it is used here to include shore-based activities. Hence, for example, a fleet may be all the purse seine vessels in a specific sardine fishery, or all the fishers setting nets from the shore in a tropical multispecies fishery.

Interested party or ***Interest group*** refers to any person or group recognized by the State or States as having a legitimate interest in the conservation and management of the resources being managed. This term is more encompassing than the term stakeholder. Generally speaking, the categories of interested parties will often be the same for many fisheries and could include contrasting interests: commercial/recreational, conservation/exploitation, artisanal/industrial, fisher/buyer-processor-trader as well as

UNIVERSIDAD DE SEVILLA
Fac. G.ª e Historia-Biblioteca

governments (local/State/national). The general public and the consumers could also be considered as interested parties in some circumstances.

Management objective is a target that is actively sought and provides a direction for management action. For example, achieving a reasonable income for individual fishers is one possible economic objective of fisheries management.

Recruits to a stock are the new age group of the population entering the exploited component of the stock for the first time or young fish growing into or otherwise entering that exploitable component.

Reference point is an estimated value derived from an agreed scientific procedure and/or an agreed model which corresponds to a state of the resource and/or of the fishery and can be used as a guide for fisheries management. Some reference points are general and applicable to many fish stocks, others should be stock-specific. A distinction should be made between target reference points and limit reference points, or thresholds, the latter representing low states of the stock to be avoided.

Species assemblage is the term used to describe the collection of species making up any co-occurring community of organisms in a given habitat or fishing ground.

Sustainable use means the use of components of biological diversity in a way and at a rate that does not lead to the long-term decline of biological diversity or of any of its components, thereby maintaining their potential to meet the needs and aspirations of present and future generations.

Table 1. Desirable data and information requirements for fisheries at the level of policy making, according to nature and use of the data

Data type			
Related to the resource	Fishery characteristics	Social and economic information	Monitoring, control and surveillance
Summary of recent landings by fishery Summary of potential yields by fishery, with options of possible alternative approaches Probable inter-annual variability in yield and any likely long-term trends in resource productivity Details on environmental constraints and sensitive habitats Details on the implications of any international agreements which affect the fisheries	Summary of types of fishery and fleet and gear characteristics for each fleet Number of fishing units for each fleet, at present Extent and importance of recreational fisheries, where applicable Key fishing grounds and their characteristics Summary of number and distribution of landing sites The impact of fishing gear and practices on the environment and on the ecosystem Details of the costs of fishery management	Summary of existing user rights systems of each fishery and fleet Major interest groups and their 'stakes', including gender and age sub-divisions within each interest group and likely policy implications Any trends influencing or likely to influence fisheries, e.g. demographic changes, political changes, migrations, etc. Employment characteristics by fishery and fleet and possible alternative sources of employment	Summary of successes or problems in monitoring and control by fishery and fleet Financial and institutional implications of different policy options for monitoring and control Details of existing arrangements and potential for partnerships or co-management with user or interest groups

Table 1 (cont'd)

Data type			
Related to the resource	Fishery characteristics	Social and economic information	Monitoring, control and surveillance
		Contributions to national or local economy by fishery and fleet Existing or likely developmental activities and their implications for fisheries Details of any subsidies being paid to fishers and estimated costs of reducing over-capacity Characteristics of and trends in markets Implications of State macro-economic policies which could influence fisheries Details on any existing international agreements on trade, cooperation, etc., which affect fisheries	

Table 1 (cont'd)

Data type			
Related to the resource	Fishery characteristics	Social and economic information	Monitoring, control and surveillance
		Existing institutional structures related to the fishery, including traditional institutions Details on any existing or possible conflicts between fisheries or fleets, including the causes	

Table 2. Desirable data and information requirements for fisheries for the formulation of management plans, according to nature and use of the data. The information required for determination of the overall fisheries policy is also relevant to the formulation of management plans

* = Desirable but lower priority

Data type			
Related to the resource	Fishery characteristics	Social and economic information	Monitoring, control and surveillance
Historical and current catch data (in weight or numbers), including directed and by-catch and discards, for fishery and fleets Size and/or length composition of catch per fleet Sex and maturity composition of catch per fleet(*) Age composition of catch per fleet(*) Time, date and locality of all catches(*) Fishery independent biomass estimates	Gear used by different fleets and knowledge of its selectivity Number of fishing units (e.g. vessels and fishers) in each fleet Numbers and localities of landing sites and fishing units operating from or landing at each site Total effort for each fleet Relative fishing power of different fishing units Area fished by each fishing unit	Description of the types of production units in the fishery and the number of each type of production unit per fleet Details of user or access rights systems related to the fishery Total number of fishers employed in all fisheries-related activities, with details on gender and age-group characteristics	Existing monitoring and control systems for the fishery and fleets within it Known strengths and weaknesses of existing systems Implications (personnel, costs, benefits, etc.) of range of approaches for monitoring and control Potential for greater user participation

Table 2 (cont'd)

Data type			
Related to the resource	Fishery characteristics	Social and economic information	Monitoring, control and surveillance
Results of stock assessments indicating potential yields and resource status under different harvesting strategies Annual estimates of number of recruits entering fishery(*) Stomach contents data for knowledge of trophic relations Data on mass of species consumed per predator type and feeding preferences of predators(*)	Detailed characteristics on equipment per vessel which could influence efficiency (e.g. GPS, echo-sounder, etc.)(*) Mass of catch by commercial size category(*) Implications for each fleet for range of management approaches Comprehensive data, per catch, on effort used, exact position, depth fished and other data relevant to characteristics of the catch for each fleet(*)	Existence of, and possible solutions to, any conflicts between fisheries or fleets Total landed value of the catch for each fleet and any other benefits Details on processing of catch and on markets, as well as benefits derived from these activities Existing or potential systems (institutions) and their potential roles in shared responsibility or co-management	Existing legislation and regulations Additional legislation and regulations, or modifications, required for range of management approaches

Table 2 (cont'd)

Data type			
Related to the resource	Fishery characteristics	Social and economic information	Monitoring, control and surveillance
Time series of indices of environmental characteristics (e.g. sea surface temperature)(*)		Details on full costs of fishing by fleet and processing, marketing and distribution costs Specific international trade or cooperation agreements relevant to fisheries Details on socio-economic characteristics of national or local non-fishing activities which do or may impinge on the fisheries Procedures for consultation and joint decision-making	

Table 3. Desirable data and information requirements for fisheries for implementation of the management plan, according to nature and use of the data. The data and information required for the formulation of the management plan are also relevant to the implementation of the management plan

Data type			
Related to the resource	Fishery characteristics	Social and economic information	Monitoring, control and surveillance
Most recent data on indices used in management procedure (e.g.commercial CPUE, estimated biomass, etc.) Information on biological or environmental features which could affect interpretation of indices Information on any unexpected event related to the stock (e.g. unusual recruitment, natural mortality, environmental conditions) which could warrant departure from management procedures	Total catch and effort data for the fishery or, if heterogenous, per fleet Unusual features of fishery or fleet behaviour which could influence interpretation of stock indices used in the management procedure Changes in fishery or fleet composition which could impact on management procedures	Unexpected social changes which could require departure from management procedure, e.g. movements, changes in patterns of access Unexpected economic changes, e.g. in markets, returns or costs which could seriously impact the management plan Social and economic performance of fisheries and fleets in relation to objectives of management plan	Name of each fisher or licensed fishing unit (e.g. vessel) Address or port of registry of each vessel or fishing unit Name and address of owner of each fishing vessel or unit Information from each fishing unit necessary for enforcing management measures (e.g. catch, effort deployed, catch position, etc.)

Table 3 (cont'd)

Data type			
Related to the resource	Fishery characteristics	Social and economic information	Monitoring, control and surveillance
Status of the stock in relation to trends anticipated in the management plan		Details on the nature and causes of any serious conflicts within the fishery	In the case of vessels: (1) date and place built (2) type of vessel (3) length of vessel (4) vessel markings (5) type of gear (6) international radio call sign Incidence and causes of any serious and ongoing violations of the management plan

Table 4. Outline of possible topics in a fishery management plan

- Title
- Area of operation of the fishery and under which jurisdiction it falls
- History of fishing and management
- Particulars of the recognized groups with interests in the fishery (interest groups)
- Details of consultations leading to formulation of the management plan
- Arrangements for on-going consultations with interest groups
- Details of decision-making process or processes, including the recognized participants
- Objectives for the fishery:
    - resource (Section 1.3.1)
    - environmental (Section 1.3.2)
    - biodiversity and ecological (Section 1.3.3)
    - technological (Section 1.4)
    - social (Section 1.5)
    - economic (Section 1.5)
- Outline of the fishery resources including particulars of life histories as appropriate
- Outline of fleet types or fishing categories participating in the fishery
- Outline of status of the stocks as indicated by stock assessments, including a description of the assessment methods, standards, and stock indicators, biological limits, etc.
- Description of the aquatic ecosystem, its status and any particularly sensitive areas or features influencing or affected by the fishery (Section 1.3.2)
- Details of non-fishery users or activities which could impact on the fishery, and arrangements for liaison and co-ordination. This may be particularly important in inland and coastal fisheries.

Table 4 (cont'd)

- Details of those individuals or groups granted rights of access to the fishery, and particulars of the nature of those rights (Sections 3.2 and 3.3)
- Description of the measures agreed upon for the regulation of fishing in order to meet the objectives within a specified time-frame (Section 3.1). These may include general and specific measures, precautionary measures, contingency plans, mechanisms for emergency decisions, etc.
- Specific constraints, e.g. details of any undesirable bycatch species, their conservation status and measures taken to reduce this as appropriate
- Details of any critical environments or sources of concern and actions required to address them
- Particulars of arrangements and responsibilities for monitoring, control and surveillance and enforcement
- Details of any planned education and training for interest groups
- Date and nature of next review and audit of the management plan

Some of the above may be of a generic nature and hence be dealt with in the general rules of fishing (e.g. a national fishery legislation), in which case these can be referred to in the plan, without repeating all the details. However, specific points or detail may be required for specific fisheries.